Conference Proceedings of the Society for Experimental Mechanics Series

Series Editor

Kristin B. Zimmerman, PhD.
Society for Experimental Mechanics, Inc.,
Bethel, CT, USA

The Conference Proceedings of the Society for Experimental Mechanics Series presents early findings and case studies from a wide range of fundamental and applied work across the broad range of fields that comprise Experimental Mechanics. Series volumes follow the principle tracks or focus topics featured in each of the Society's two annual conferences: IMAC, A Conference and Exposition on Structural Dynamics, and the Society's Annual Conference & Exposition and will address critical areas of interest to researchers and design engineers working in all areas of Structural Dynamics, Solid Mechanics and Materials Research.

More information about this series at http://www.springer.com/series/8922

Alireza Amirkhizi • Jacob Notbohm • Nikhil Karanjgaokar
Frank W DelRio
Editors

Challenges in Mechanics of Time Dependent Materials, Mechanics of Biological Systems and Materials & Micro-and Nanomechanics, Volume 2

Proceedings of the 2021 Annual Conference & Exposition on Experimental and Applied Mechanics

Springer

Editors
Alireza Amirkhizi
UML North Campus, Dandeneau Hall
University of Massachusetts Lowell
Lowell, MA, USA

Jacob Notbohm
University of Wisconsin–Madison
Madison, WI, USA

Nikhil Karanjgaokar
Higgins Lab
Worchester Polytech Institute
Worcester, MA, USA

Frank W DelRio
Material Measurement Laboratory
National Institute of Standards and Technology
Boulder, CO, USA

ISSN 2191-5644 ISSN 2191-5652 (electronic)
Conference Proceedings of the Society for Experimental Mechanics Series
ISBN 978-3-030-86736-2 ISBN 978-3-030-86737-9 (eBook)
https://doi.org/10.1007/978-3-030-86737-9

This Springer imprint is published by the registered company Springer Nature Switzerland AG
The registered company address is: Gewerbestrasse 11, 6330 Cham, Switzerland

Preface

Challenges in Mechanics of Time-Dependent Materials, Mechanics of Biological Systems and Materials, and Micro-and Nanomechanics represents one of four volumes of technical papers presented at the 2021 SEM Annual Conference & Exposition on Experimental and Applied Mechanics organized by the Society for Experimental Mechanics held on June 14–17, 2021. The complete Proceedings also includes volumes on *Dynamic Behavior of Materials; Mechanics of Composite, Hybrid & Multifunctional Materials, and Fracture, Fatigue, Failure and Damage Evolution; Thermomechanics & Infrared Imaging, Inverse Problem Methodologies, Mechanics of Additive & Advanced Manufactured Materials;* and *Advancement of Optical Methods & Digital Image Correlation.*

Each collection presents early findings from experimental and computational investigations on an important area within Experimental Mechanics, the Mechanics of Time-Dependent Materials, Fracture, Fatigue, Failure, and Damage Evolution being some of these areas.

The Time-Dependent Materials track was organized to address constitutive, time (or rate)-dependent constitutive, and fracture/failure behavior of a broad range of materials systems, including prominent research in progress in both experimental and applied mechanics. Papers concentrating on both modeling and experimental aspects of Time-Dependent Materials are included.

The Biological Systems and Materials segment of this volume summarizes the exchange of ideas and information among scientists and engineers involved in the research and analysis of how mechanical loads interact with the structure, properties, and function of living organisms and their tissues. The scope includes experimental, imaging, numerical, and mathematical techniques and tools spanning various length and time scales. Establishing this symposium at the Annual Meeting of the Society for Experimental Mechanics provides a venue where state-of-the-art experimental methods can be leveraged in the study of biological and bio-inspired materials, traumatic brain injury, cell mechanics, and biomechanics in general. A major goal of the symposium was for participants to collaborate in the asking of fundamental questions and the development of new techniques to address bio-inspired problems in society, human health, and the natural world. The 2021 Symposium is the 11th International Symposium on the Mechanics of Biological Systems and Materials. The organizers would like to thank all the speakers and staff at SEM for enabling a successful program.

The Micro- and Nanomechanics segment of this volume focuses on specialized scientific areas that involve miniaturizing conventional scale components and systems to take advantage of reduced size and weight and/or enhanced performance or novel functionality. These fields also encompass the application of principles ranging from the micron scale down to individual atoms. Sometimes these principles borrow from conventional scale laws but often involve new physical and/or chemical phenomena that require new behavioral laws and impart new properties to exploit. Studying how mechanical loads interact with components of these scales is important in developing new applications, as well as assessing their reliability and functionality. Establishing this symposium at the Annual Meeting of the Society for Experimental Mechanics provides a venue where state-of-the-art experimental methods can be leveraged in these endeavors.

The 2021 Symposium is the 22nd in the series and addresses pertinent issues relating to design, analysis, fabrication, testing, optimization, and applications of micro- and nanomechanics, especially as these issues relate to experimental mechanics of microscale and nanoscale structures.

The track organizers thank the presenters, authors, and session chairs for their participation and contribution to these tracks. The support and assistance from the SEM staff is also greatly appreciated.

Lowell, MA, USA
Madison, WI, USA
Worcester, MA, USA
Boulder, CO, USA

Alireza Amirkhizi
Jacob Notbohm
Nikhil Karanjgaokar
Frank DelRio

Contents

Chapter 1
Advance of Collaborative Twinning Fields in Magnesium AZ31 via the Strain and Residual Intensity Channels in Microscopic Image Correlation

Necdet A. Özdür and C. Can Aydıner

Abstract Profuse mechanical twinning is a central element in the mechanical behavior of Magnesium alloys. In the attempt to further employ this lightest class of structural metals in engineering applications, this micro-mechanism underlies significant challenges due to its abrupt and unipolar activity. A significant trait in mechanical twinning is autocatalytic activity, crudely, twinning in one grain triggering the twin in the next. In a sharply textured Magnesium polycrystal, this entails a collaborative twinning front that abruptly advances across a vast number of grains, producing a Lüders-like banded deformation. An area-scanning version of DIC with optical microscopy has been instrumental in studying this deformation regime in situ, by producing quantitative strain data that bridges grain and sample length scales. These measurements elucidate the multi-scale nature of the bands, clearly showing that a sample-scale collaborative twin band is composed of a fine mesh of microscopic strain localization bands at the grain scale. To construct the physical connection of this strain localization mesh to the underlying collaborative twinning field, however, one typically requires a separate microscopy study where the twins are observed with electron microscopy. Recently, we have shown that a detailed microscopic image of the twin band field can be derived from the same images that are recorded for DIC analysis. This is achieved via an imaging mode called residual intensity that refers to a material-point-aligned subtraction of reference and deformed images—only the latter containing the local intensity offsets imposed by the twin bands. The result is a pixel-resolution imagery of the twins that can be superimposed over the calculated DIC strain fields. To maximize the benefits of this approach, DIC is conducted with high resolution optics and a maximal magnification for optical DIC with ~250 data points per grain (average grain size 13 μm).

In the current effort, we will make a more detailed analysis of the band boundary and its advance, similarly utilizing strain and residual intensity fields to obtain combined deformation and morphology data. As observed over residual intensity maps, the macroscopic band boundary comprises a comb-like array of twins that protrude into the dormant material. We detail the deformation/twin fields as the band advances normal to this boundary, namely, when the next section of the dormant material undergoes collaborative twinning.

Key words Magnesium · Twinning · Microscopy · Image correlation · Residual intensity

1.1 Introduction

The mechanical behavior of wrought Magnesium is highly inhomogeneous and anisotropic due to its sharp crystallographic textures combined with the variety and activation ease of Magnesium deformation micro-mechanisms. This leads to strong load path dependence for these materials, namely, the set of micro-mechanisms that accommodate strain vary with the shape and sense of the strain imposed [1–3]. The most fundamental changes in these deformation regimes are due to deformation twin mechanisms, which are further unipolar. This means if a twin is profusely activated along one load path (e.g., compression), it is not activated over the reverse path (e.g., tension) [2, 4, 5]. The governing set of micro-mechanisms that accommodate strain also change the spatial signature of how the strain is accommodated and the strain heterogeneity levels. Coordinated (autocatalytic) twin activity [6–8] where twins propagate by triggered events lead to pronounced strain heterogeneity in the form of multi-scale shear bands [1, 9].

Quantifying strain heterogeneity requires spatially-resolved strain measurement and the standard tool for this purpose has digital image correlation (DIC) [10]. DIC can be applied at multiple length scales and it has been instrumental for investigating

N. A. Özdür · C. C. Aydıner (✉)
Department of Mechanical Engineering, Boğaziçi University, Istanbul, Turkey
e-mail: necdet.ozdur@boun.edu.tr; can.aydiner@boun.edu.tr

A. Amirkhizi et al. (eds.), *Challenges in Mechanics of Time Dependent Materials, Mechanics of Biological Systems and Materials & Micro-and Nanomechanics, Volume 2*, Conference Proceedings of the Society for Experimental Mechanics Series,
https://doi.org/10.1007/978-3-030-86737-9_1

wrought Magnesium strain fields. In situ microscopic implementations of DIC are used to resolve the fine details of deformation heterogeneity inside macroscopic twin bands with statistically significant data sets. These help illuminate the collaborative nature of individual twinning events [1, 2, 9].

Strain information provided by DIC, however, does not directly offer any conclusive evidence of the underlying micro-mechanisms. That means, parallel microscopy efforts are typically necessary to ascertain that the twins underlie the strain localizations. On the other hand, at magnifications that are high enough to resolve strain localizations due to individual twins, the DIC images are affected by the deformation steps caused by the twin activity on the surface. This means a local contrast will be introduced to the recorded images over the twin band [1]. Recently, we have proposed a novel method to single out and monitor these effects on the surface just by using the DIC images [11]. Here, a DIC-derived parameter "residual intensity" is posed as an in situ mapper of new twin activity. Residual intensity is a material-point aligned subtraction of images recorded at reference and deformed configurations. This is a separate, pixel-resolution data channel that naturally complements the deformation measurements obtained from DIC analysis that are lower resolution by about an order. With its higher resolution, and as a parameter that solely focuses on deformation-mediated image contrast changes, residual intensity can identify morphological shapes of deformation micro-mechanisms, which would otherwise require meticulous SEM analyses to detect. Of course, parallel microscopy by SEM is further limiting in the context of in situ area-scanning implementation of DIC, since SEM can only be conducted ex situ at a single data point (typically chosen as the final point). Conversely, residual intensity is available at any data point. The effectiveness of residual intensity maps for twinning Magnesium was demonstrated over a rolled sample of Magnesium AZ31 under compression, where the interwoven twin fabric that constitutes the overall macroscopic twin band, and its jagged extensions protruding toward the dormant material is imaged with extreme detail [11].

Residual intensity ideally removes all baseline surface features that are shared between reference and the deformed images (including the initial DIC pattern). Therefore, the residual intensity map obtained from a set of images is expected to contain only the "relative, deformation-induced, changes in surface features" between the two states of DIC analysis. Unlike classical microscopy modes that allow imaging just at a single moment in time, residual intensity provides a unique view into the relative changes between two arbitrary points in time. One of the advantages of this methodology is that it can allow a time-resolved monitoring of the deformation structures when DIC is conducted among consecutive loads. In this application, residual intensity should solely focus on the deformation agents that formed in this particular load increment. The goal of this study is then to showcase this capability. To this end, we will introduce an extra load step over the already-twinned Magnesium sample in Özdür et al. [11] and observe the further propagation of the twin band onto the neighboring regions with DIC analyses that compare the extra load step with the previous one. We anticipate such a relative analysis to produce strain and residual intensity maps that highlight formations only in the expansion zone of the shear band.

1.2 Materials and Methods

This study considers an additional load step over the Magnesium AZ31 sample that is considered in Özdür et al. [11]. Only basic details about the sample are recovered in this section, and the reader is referred to this effort for more detailed information on the sample preparation, crystallographic texture, and experimental procedures.

The dog-bone sample of Magnesium was electric-discharge machined from a rolled plate such that the DIC observation surface is aligned with the RD-ND plane and the loading axis is aligned with the RD as indicated in Fig. 1.1a (rolling, normal, transverse directions are aptly abbreviated as RD, ND, TD here). The resulting rolling texture favors activation of $\langle 10\bar{1}2 \rangle$ and $\{10\bar{1}\bar{1}\}$ tensile twin pair when the sample is under compression along RD, which is indicated on the sketch of a unit cell with statistically the most common orientation in Fig. 1.1b.

The observation surface is first metallographically prepared to reveal grain boundaries and then added a sparse layer of paint. The sample is then loaded in compression to a near perfectly plastic region called the twin plateau over the stress strain curve [2, 9]. There is a very detailed multi-scale presentation of the state of the sample at the previous load point of this sample in Özdür et al. [11]. Here, we inherit the sample in this state and impose an additional 0.45% (sample-averaged) strain to expand the twin-mediated shear bands. Unfortunately, there is more than 2 months of time delay between the previous load point and the additional procedure of the current effort. This time delay created superfluous contrast on the surface already with environmental effects such as oxide formations and dust accumulation. These naturally contaminate the residual intensity channel (but not to an extent to completely suppress its twinning-signal) as will be discussed later.

The experimental setup shown in Fig. 1.1c is mainly comprised of stationary macroscopic and microscopic optical lines for macro-DIC (FLIR Grasshopper 3 camera, 2.3 MP Sony IMX174 monochrome sensor, Edmund Optics 0.5× telecentric lens) at 11.6 μm/pixel resolution, and micro-DIC (FLIR Grasshopper 3 camera, 5 MP Sony IMX250 monochrome sensor, Navitar

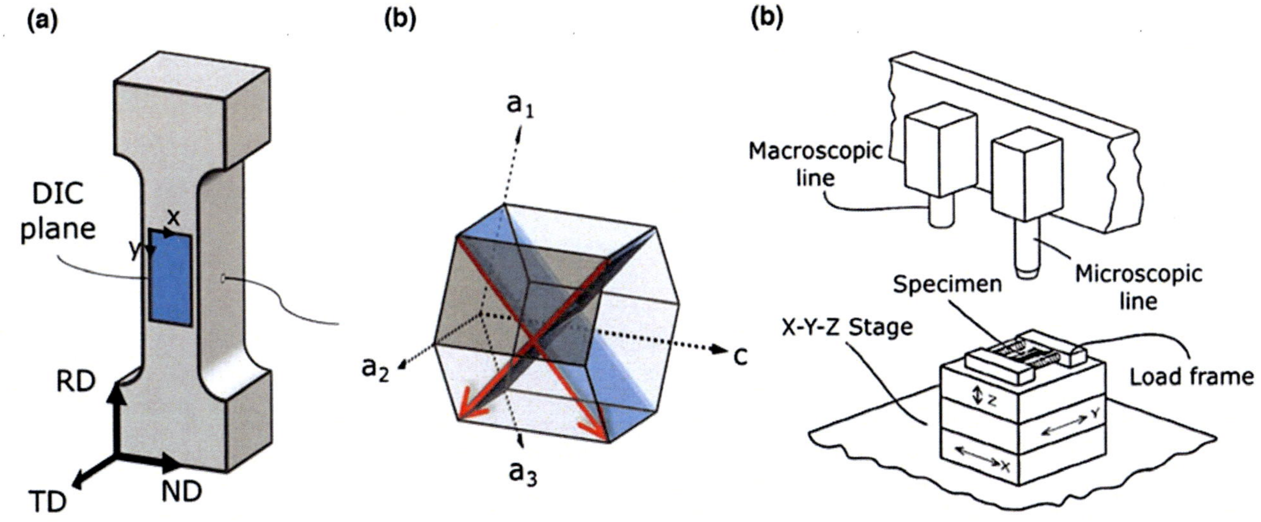

Fig. 1.1 Sketches of **(a)** the dog-bone sample and its alignment with respect to the rolled plate, **(b)** the most probable orientation of the HCP unit cell within the sample, **(c)** experimental hardware

Ultrazoom 6000 tube lens, Mitutoyo M Plan APO 20× objective) at 0.0799 μm/pixel resolution; and a small form factor tension-compression load frame (10 kN, Kammrath & Weiss) that is mounted horizontally on top of an *X-Y-Z* positioning stage (Newport M-ILS series of *X-Y*, GTS30V for *Z* positioner). The entire setup is positioned on a vibration isolation table (Newport RS4000). An in-house instrument software is used to facilitate the automated coordination of cameras and the motion of the stages.

Standard experimental procedure employs continuous monitoring with the macro camera (and live macro-DIC analysis) as the sample is loaded. When significant deformation changes are observed, the sample is sent under the micro-DIC line by the automated script and micro-DIC area-scans are conducted. These record images of the surface over a predefined grid of overlapping frames. In this experiment, we inspect a 3 frame by 3 frame (3 × 3) sub-region of the larger 7 × 5 area-scan of the previous work [11]. The 3 × 3 frame grid produces a combination image of 30 Megapixels, covering a region of 0.40 × 0.48 mm^2.

Each frame in the automated micro-DIC scan is auto-focused and auto-located (material field correction) in real time as corrective measures. Auto-focusing counteracts the defocusing effect of expansion in *Z*-direction as the sample is deformed in compression. This is achieved by automatic selection of the image with best focus measure value [1] from an image stack that is recorded with fine increments in *Z*-direction. On the other hand, material field correction counters the overall shift in *X-Y* plane of each frame in deformed configuration with respect to its counterpart in reference configuration. To do this, it measures the shift via an FFT-based cross-correlation algorithm. Since each frame in the deformed grid is analyzed with respect to its counterpart in the reference configuration, material field correction is crucial to ensure that the bulk of material points recorded in the reference frame are recorded in the corresponding deformed frame. This measure thus guarantees a continuous data field is presented when deformation data from all frames are stitched together.

Here, residual intensity is briefly defined in conjunction to the DIC basics (once more leaving details to the primary effort [11]). DIC is used to obtain a linear mapping function, χ, that matches each pixel coordinate in reference configuration, $\mathbf{x}$, to a unique pixel coordinate in deformed configuration, $\mathbf{y}$, such that:

$$y = x + u(\mathbf{x}) = \chi(\mathbf{x}) \tag{1.1}$$

Here $u(\mathbf{x})$ denotes the material point displacement vector at every pixel coordinate. An ideal mapping χ with known displacements should match the same material point at the reference image $f(\mathbf{x})$, to its displaced counterpart in the deformed image $g(\mathbf{y})$ such that $f(\mathbf{x}) = g(\chi(\mathbf{x}))$. However, even with a perfect mapping, the observed light intensity of material points changes between reference and deformed images due to multiple error sources. This intensity differential is here called "residual intensity" defined as:

$$\Delta I = f(\mathbf{x}) - g(\chi(\mathbf{x})) = f(\mathbf{x}) - \widetilde{g}(\mathbf{x}) \tag{1.2}$$

Here, $\widetilde{g}(\mathbf{x})$ is the light intensity field recorded at the deformed surface mapped back to the reference (material) coordinates. As will be noted shortly, the additional step (compared to a conventional DIC application) in making a per-pixel calculation of ΔI is spatial interpolation of the displacement fields. Incidentally, the main principle of DIC algorithm is to minimize a suitable norm of the residual intensity ΔI:

$$\min_{u} \sum [f(\mathbf{x}) - g(\mathbf{x} + u(\mathbf{x}))]^2 \tag{1.3}$$

Here, intensities are zero-normalized; it is not shown in the above equations for brevity. Note that calculation of displacements in Eq. 1.3 takes place over a sparse data grid $\mathbf{x}_{i0}$, instead of the full grid of image coordinates to save computational resources. Calculated displacements are upscaled from DIC grid resolution $u(\mathbf{x}_{i0})$ to the entire image domain $u(\mathbf{x})$ by interpolation. Similarly, intensity values of $\widetilde{g}(\mathbf{x})$ that do not lie on integer pixel coordinates can be interpolated from the surrounding known intensity fields. Özdür et al. shows that the order of interpolation used here has no obvious effect in obtaining accurate residual intensity maps, provided that the DIC displacements are calculated within reasonable accuracy; because unlike strain maps, residual intensity requires accuracy in the displacements themselves, not in their derivatives [11].

The DIC analysis in this work employs a subset-based, inverse-compositional algorithm in the minimization process as introduced in [12]. (Note our previous effort [11] used an Augmented Lagrangian DIC algorithm [13]. As we show here, residual intensity concept remains viable with a subset-based algorithm as well.) 41 × 41 pixel subsets are utilized all through (3.28 × 3.28 μm^2 and 0.47 × 0.47 mm^2 for micro- and macro-DIC, respectively) over a data grid $\mathbf{x}_{i0}$ with 10-pixel spacing. This corresponds to ~44,000 data points per frame and ~390,000 data points in the entire area-scan. Stitching of the data fields in the micro-DIC area-scan is implemented with the "grid-collection stitching" plugin in ImageJ [14–16]. Displacement derivatives are estimated by filtering the DIC displacement results with a 3 × 3 Savitzky-Golay differentiator kernel [17]. Displacements are upscaled to image resolution using bicubic interpolation when producing ΔI maps. All scripts used in this analysis are written purely in Python [18], with the use of additional functionality from SciPy [19], NumPy [20, 21], Matplotlib [22], and scikit-image [23].

1.3 Results and Discussion

Figure 1.2a–c shows macro-DIC images of the sample for all load states under inspection: Fig. 1.2a shows the *undeformed* macro-DIC image, Fig. 1.2b is the macro-DIC image of the *previous load point*, and Fig. 1.2c is the macro-DIC image of the *current load point*. Here, the *previous load point* refers to the load state attained and deeply investigated in the previous study [11], while the current load point denotes the load step we introduced in this work. There are two sets of DIC analyses in this effort that incorporates these three states. The first is conducted between the undeformed configuration (the state in Fig. 1.2a) and the previous load point (the state in Fig. 1.2b). The second is a *relative* DIC analysis of previous load point (the state in Fig. 1.2b) and the current load point (the state in Fig. 1.2c) that we introduce in this study. We will use the superscript "rel" to differentiate the parameters of this second-type analysis all through the text, where it stands for "relative." All strain and residual intensity maps in the following figures are specifically plotted at the *previous* load configuration (the state in Fig. 1.2b) for both types of analysis. This means the strains/features that formed up to *previous* load configuration and the additional strains/features that form afterwards can be overlaid in a common coordinate system.

Note the formation and advancement of two macro-scale twin bands become faintly visible in the macro-images of Fig. 1.2. The undeformed image in Fig. 1.2a has no obvious feature, while two separate bands create a visible contrast on the image of the previous load point in Fig. 1.2b, and these bands thicken further at the current load point in Fig. 1.2c. The advancement of the macro-scale band is clearer in the strain maps obtained from the region enclosed with yellow dotted lines.

The macro-DIC strain maps are given in Fig. 1.2d, e. Figure 1.2d shows the transverse strain (ε_{xx}) and the axial strain (ε_{yy}) maps obtained from a macro-DIC analysis between undeformed configuration and the previous load point. The macroscopic band region, the boundary of which is crudely emphasized with dashed black lines, exhibits a nominal −2% strain in *y*-direction and 2% strain in *x*-direction. The remaining part of the material appears to be dormant. Conversely, Fig. 1.2e shows the strain maps ε_{xx}^{rel} and ε_{yy}^{rel} obtained from a relative macro-DIC analysis between previous and the current load point. This time, the already-banded regions stay undisturbed, but the region right next to the dashed lines exhibit a similar level of strains, which indicates an advancement of the macroscopic bands in the direction of the black arrows. We wish to emphasize that the strain maps in Fig. 1.2e show relative strains of the current load point over the previous load point. There is no relative deformation over the already-banded region and that simply means that those regions accommodate the same strain in their

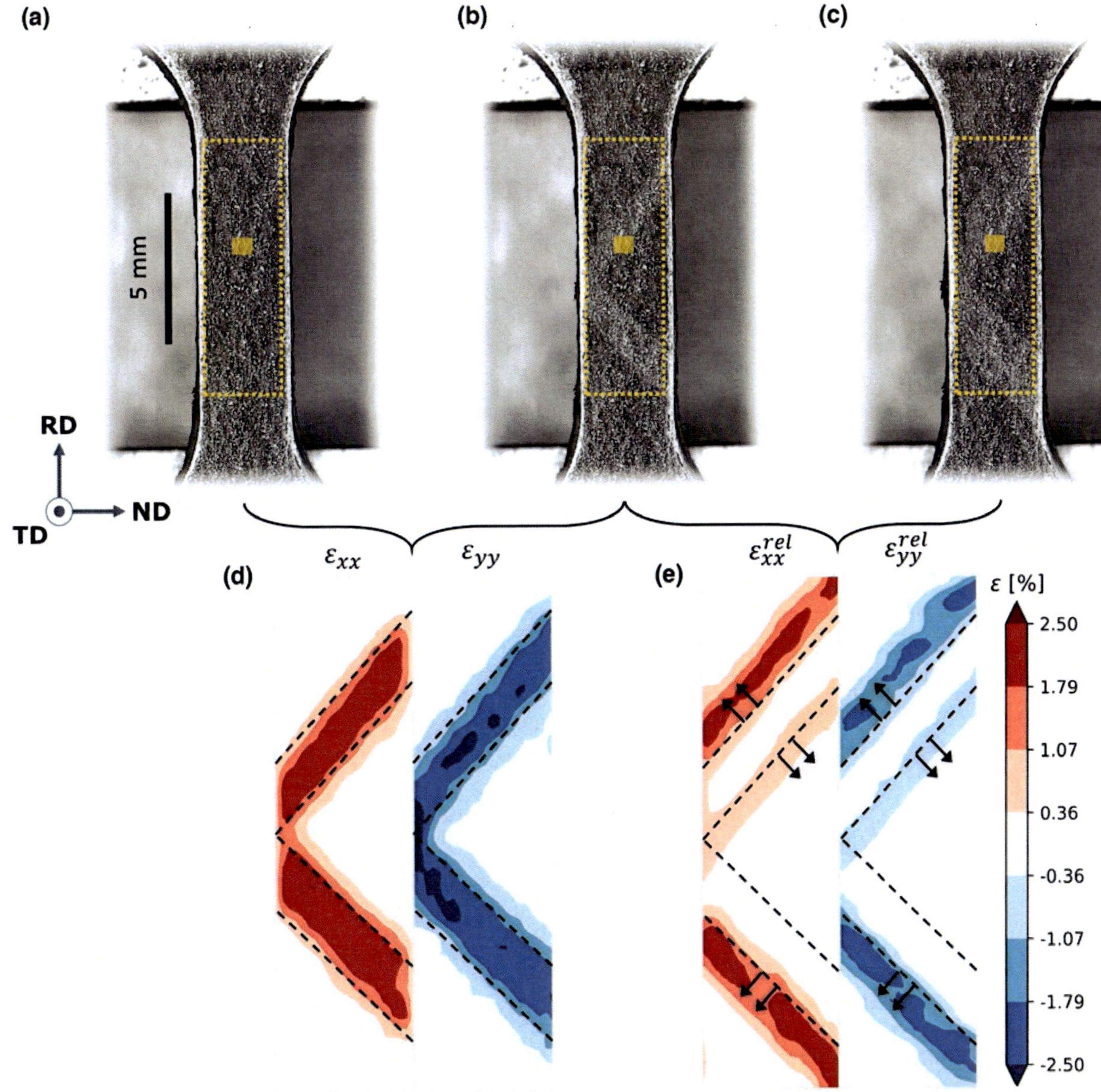

Fig. 1.2 Macro-DIC images at (**a**) undeformed configuration, (**b**) previous load point, (**c**) current load point. ε_{xx} and ε_{yy} maps obtained from (**d**) macro-DIC analysis of (**a**) and (**b**); (**e**) relative macro-DIC analysis of (**b**) and (**c**)

previous state. This behavior was previously displayed through another relative DIC analysis by Shafaghi et al. [9], i.e., the already-banded regions do not incur further strain in this deformation regime of rolled Magnesium sample.

It is also worth to mention that the two macroscopic twin bands in Fig. 1.2 seem to prefer to expand "away" from each other, while the region between the two shows a relatively limited advance. Our micro-DIC analysis region, shown with small yellow rectangles in Fig. 1.2a–c, happens to be in this sector. This region was selected to be at the shear band boundary, to better portray the advancement of the twin front.

Strain and rotation maps of micro-DIC and relative micro-DIC analyses are presented in Fig. 1.3a, b, respectively. Both rotation maps are normalized with respect to the band-average to emphasize relative the sense of rotation within the band. Strain maps in Fig. 1.3a show a highly heterogeneous deformation activity within the initial band, while the rest of the material stays mostly dormant. A comb-like pattern perpendicular to the band boundary (appears blue in rotation maps) protrudes from the initial band to the not-yet-disturbed region. ε_{xx}^{rel}, ε_{yy}^{rel}, and ω_{xy}^{rel} maps in Fig. 1.3b indicate that the band indeed advances in the direction shown by black arrows. The relative deformation inside the band extension shows an intertwined mesh of heterogeneous shear bands, but the initial band region is not disturbed between the previous and the current load points. In

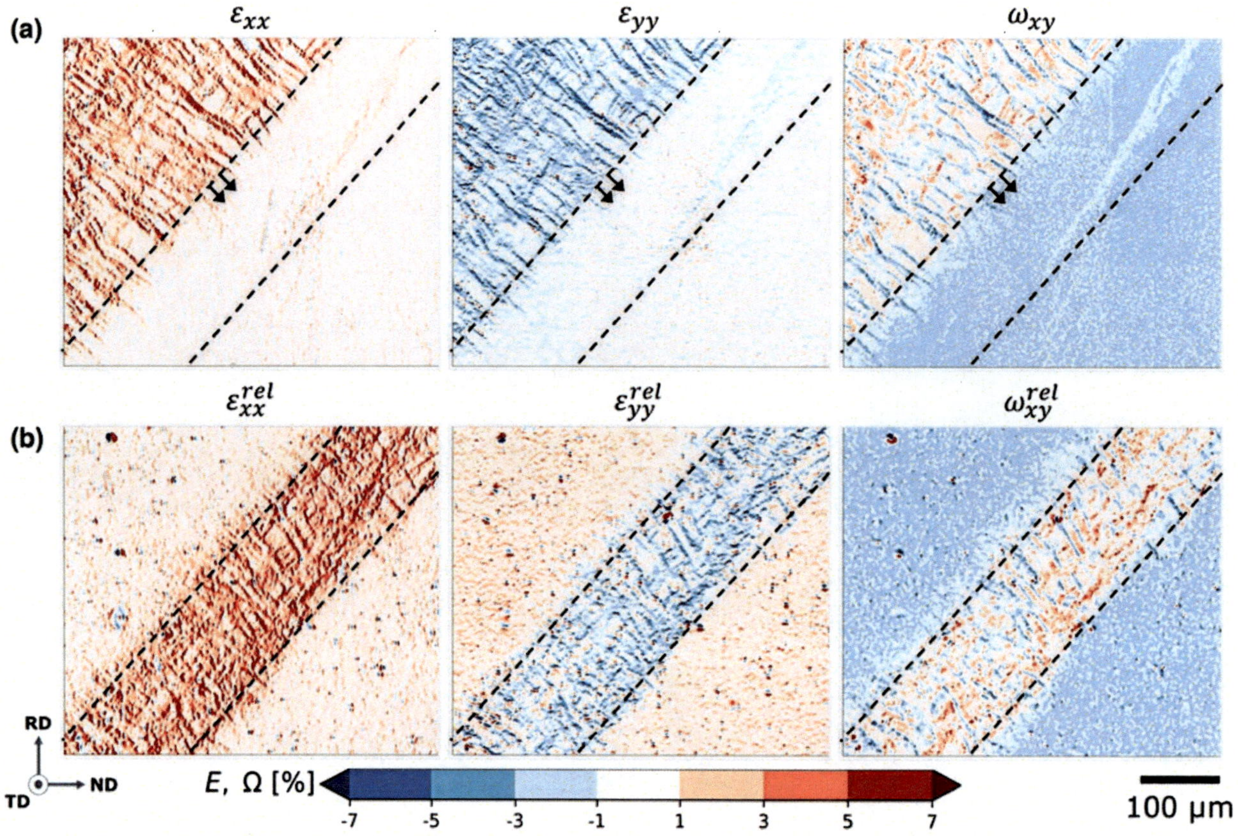

Fig. 1.3 Strain ε_{xx}, ε_{yy}, and rotation ω_{xy} maps obtained from (**a**) micro-DIC, (**b**) relative micro-DIC analysis

fact, without Fig. 1.3a, the initial band region would be totally indistinguishable from the dormant material. This again simply says that the added strain is virtually accommodated by band extension and not by further strain addition into the already-deformed material. From the micro-mechanism point-of-view, this should entail that new collaborative twin activity prefers to take place over the neighboring (dormant) material and avoids the already-twinned fields. Thanks to residual intensity, we can now evaluate this supposition directly (attempted next).

Figure 1.4a, b shows the stitched relative micro-DIC fields of previous-load and current-load images mapped at previous-load coordinates, respectively. The corresponding relative residual intensity image, $\Delta I^{rel}(x)$ in Fig. 1.4c is obtained by simple subtraction of $\tilde{g}(\mathbf{x})$ from $f(\mathbf{x})$ as explained in Eq. 1.2. The range of ΔI^{rel} is clipped from [−255, 255] to [−60, 60] gray-level to increase the contrast of its features. However, this map is not as clear as the one presented in [11], because of the complications due to the long wait-period of the sample. Specifically, opaque dust particles (that appear black in $\tilde{g}$) contaminate ΔI^{rel} field with white dots, and occasional dark patches (probably resulting from oxidation) can be seen throughout the image through their varying effect over different crystallites. Even still, the twin fabric within the band advancement region, similarly enclosed by dashed lines, can easily be discerned with much finer detail compared to the strain and rotation maps presented in Fig. 1.4. Furthermore, in comparison with the Fig. 1.5a that shows the ΔI combination figure obtained from the micro-DIC analysis of the previous experiment, ΔI^{rel} manages to suppress existing lamellar structures inside the initial macroscopic twin band and displays only the newly occurring twin instances. For example, note the pronounced twins that became visible in the already-banded region marked with green arrows in Fig. 1.4a, b. There is no sign of these structures in Fig. 1.4c, attesting to the proper elimination of the twin structures that belong to the previous state. The large lack of new instances of such lamellar structures in the already-banded region (exceptions are few and close to the band border) is the targeted morphological evidence, namely, there is by and large no further twinning activity inside the initial twin band. Rather, the incremental twin activity occurs heavily in the expansion zone.

In Fig. 1.5b, we overlay some of the localized shear bands of the (current load-previous load) relative fields on top of the previous-load ΔI field. This intends to help us identify spatially coordinated-twinning activity, basically, how the new twins come in correlation with the location of the previous twins at the shear band boundary. This overlay is produced by careful thresholding of the ω_{xy}^{rel} map in Fig. 1.3b to separate ±45° shear localizations that are parallel to the band boundary (shown in

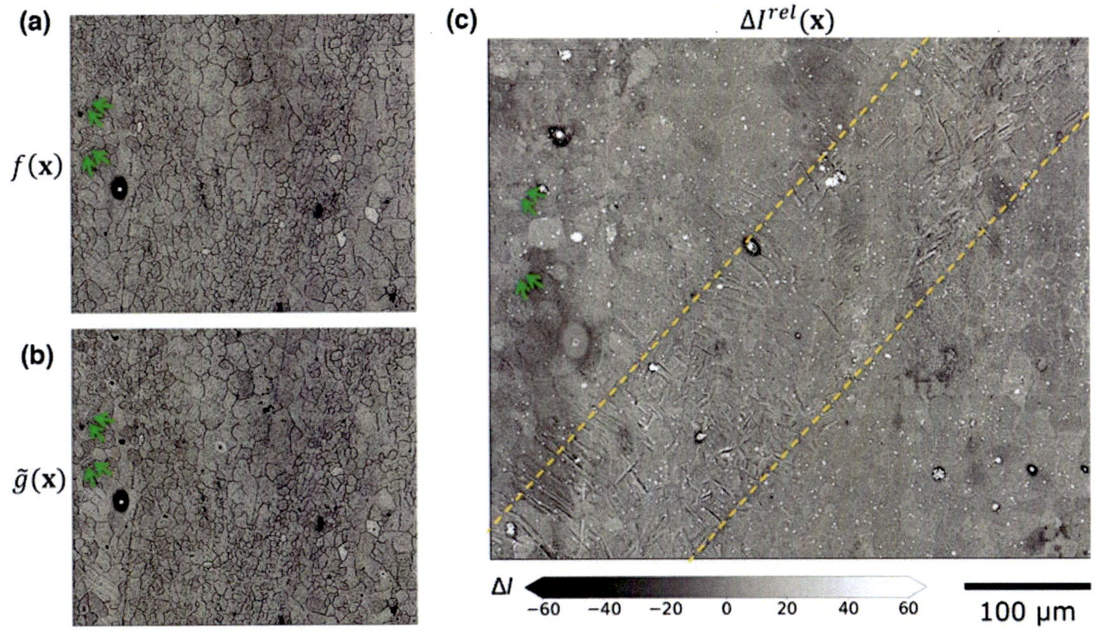

Fig. 1.4 (**a**) Previous load and (**b**) current load combined micro-DIC images. (**c**) Relative residual intensity field, ΔI^{rel}

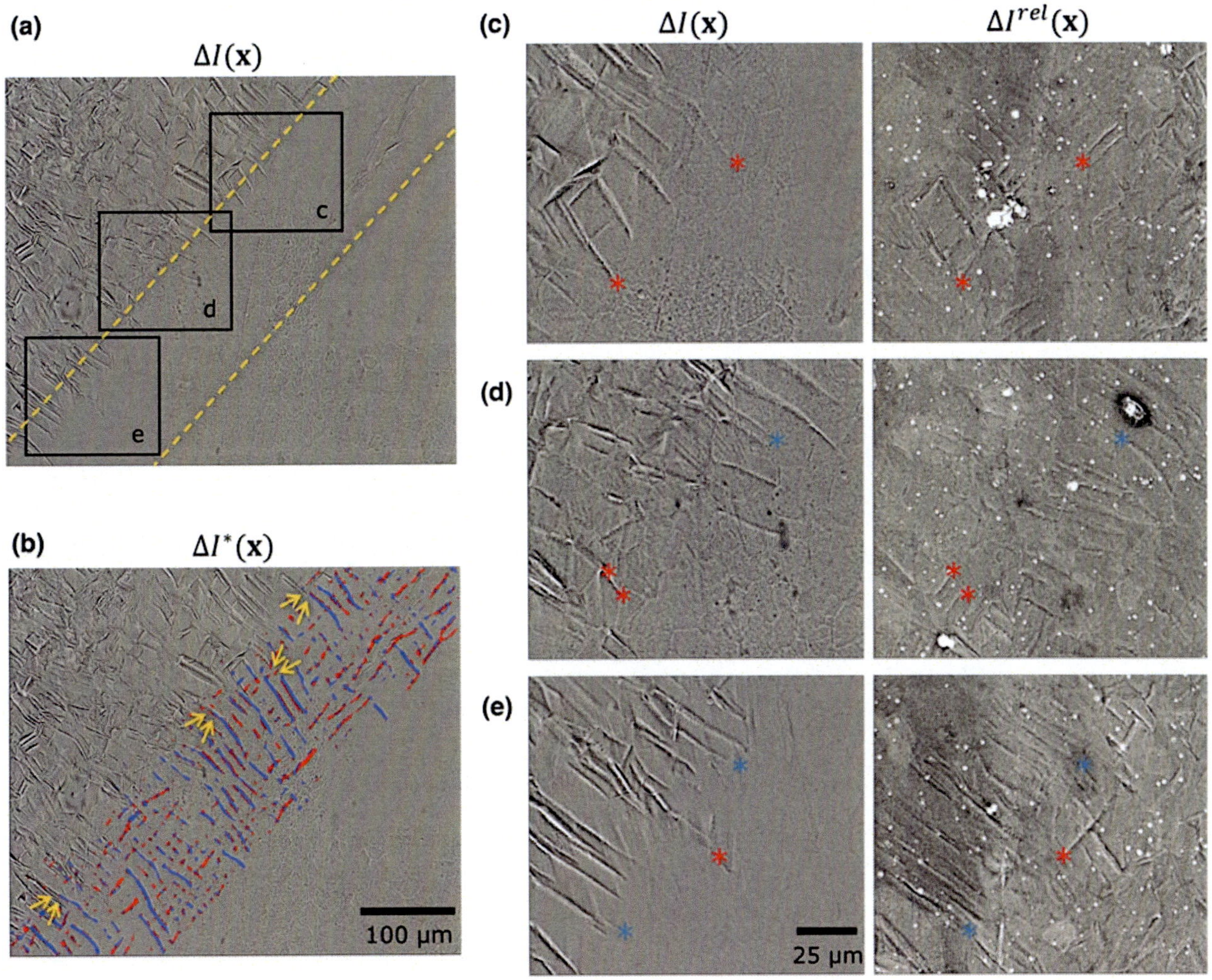

Fig. 1.5 (**a**) ΔI field produced from the micro-DIC analysis of undeformed configuration and the previous load point, (**b**) same figure as (**a**), but with an overlay of shear bands. (**c–e**) Zoomed-in ΔI and ΔI^{rel} fields of black square regions as labeled in (**a**)

red) and that are perpendicular to the band boundary (shown in blue) from each other. Yellow arrows indicate new formations of shear localizations (blue lines) from the tip of the existing twins (visible signatures in ΔI, identified as twins in [11]) at their touching points. For a close inspection, we zoom in three different regions on ΔI and ΔI^{rel} field as shown in Fig. 1.5c–e. Where these zoom-ins correspond to in the combined field can be seen in aptly labeled black rectangles in Fig. 1.5a. On these zoomed-in residual intensity fields, junctions of existing twins extending toward the dormant region on ΔI map and new formations of lamellar plasticity agents, (which spread predominantly in ±45° orientations) on ΔI^{rel} map are marked with an asterisk. Following our color map convention of rotation fields in Fig. 1.3, we used blue asterisks to indicate new initiations of twins that are perpendicular to the macroscopic twin band boundary, and red asterisks for instances where the propagation of the new twin is parallel to the band boundary. These cases of new twins emanating from the tips of existing twin formations in the previous load shows an essential micro-scale coordination in of expansion of the macroscopic twin band. One might postulate that this triggered-activity is why twin bands prefer to expand rather than form over an independent segment elsewhere over the gage section.

1.4 Conclusion

Residual intensity is a novel imaging method tied to the DIC method that accentuates changes in surface signatures of a material in two discrete points in time that correspond to reference and deformed configurations. If these changes are due to material deformation, one can get a map of plasticity agents that are introduced in this time increment. This, however, does not have to be the gross period between a deformed point and the undeformed reference. The concept can be applied among consecutive load points for better time resolution, where the ensuing residual intensity maps single out formations that occur solely over the time increment between the two load points. Note the attained time resolution is quite different from conventional microscopy tools that can investigate material states at a single point in time. Even if images are recorded with time resolution in regular microscopy, these would not be difference images that single out events over time intervals.

To this end, we utilize the novel residual intensity method in conjunction with two DIC analyses conducted over three consecutive load states to inspect the development of the macroscopic twin band in rolled Magnesium. In both macroscopic and microscopic DIC analyses of subsequent load points, we observe no new twin activity inside the already-banded region of the previous load point, but new instances of twins follow the macroscopic twin front. We further demonstrate the autocatalytic nature of the twins at the granular scales in high resolution images offered by the residual intensity fields., The relative residual intensity maps show that the propagation of the macroscopic band over the neighboring fields ensues from a coordinated twin activity of the earlier twin formations that reach out of band boundary in a comb-like pattern, which triggers formation of new twins from their tips.

Acknowledgments This work was supported by the Scientific and Technological Research Council of Turkey (TÜBİTAK), Grant No: 114M215.

References

1. Aydıner, C.C., Telemez, M.A.: Multiscale deformation heterogeneity in twinning magnesium investigated with in situ image correlation. Int. J. Plast. **56**, 203–218 (2014). https://doi.org/10.1016/j.ijplas.2013.12.001
2. Kapan, E., Shafaghi, N., Uçar, S., Aydıner, C.C.: Texture-dependent character of strain heterogeneity in magnesium AZ31 under reversed loading. Mater. Sci. Eng. A. **684**, 706–711 (2017). https://doi.org/10.1016/j.msea.2016.12.085
3. Üçel, İ.B., Kapan, E., Türkoğlu, O., Aydıner, C.C.: In situ investigation of strain heterogeneity and microstructural shear bands in rolled magnesium AZ31. Int. J. Plast. **118**, 233–251 (2019). https://doi.org/10.1016/j.ijplas.2019.02.008
4. Mohr, D., Chevin, M.-A., Greve, L.: Deformation behavior of magnesium extrusions with strong basal texture: experiments and modeling. J. Appl. Mech. **80**, 061002 (2013). https://doi.org/10.1115/1.4023958
5. Wu, L., Agnew, S.R., Ren, Y., et al.: The effects of texture and extension twinning on the low-cycle fatigue behavior of a rolled magnesium alloy, AZ31B. Mater. Sci. Eng. A. **527**, 7057–7067 (2010). https://doi.org/10.1016/j.msea.2010.07.047
6. Wang, F., Sandlöbes, S., Diehl, M., et al.: In situ observation of collective grain-scale mechanics in Mg and Mg–rare earth alloys. Acta Mater. **80**, 77–93 (2014). https://doi.org/10.1016/j.actamat.2014.07.048
7. Beyerlein, I.J.J., Capolungo, L., Marshall, P.E.E., et al.: Statistical analyses of deformation twinning in magnesium. Philos. Mag. **90**, 2161–2190 (2010). https://doi.org/10.1080/14786431003630835
8. Barnett, M.R., Nave, M.D., Ghaderi, A.: Yield point elongation due to twinning in a magnesium alloy. Acta Mater. **60**, 1433–1443 (2012). https://doi.org/10.1016/j.actamat.2011.11.022
9. Shafaghi, N., Kapan, E., Aydıner, C.C.: Cyclic strain heterogeneity and damage formation in rolled magnesium via in situ microscopic image correlation. Exp. Mech. **60**, 735–751 (2020). https://doi.org/10.1007/s11340-020-00612-6

10. Schreier, H., Orteu, J.J., Sutton, M.A.: Image Correlation for Shape, Motion and Deformation Measurements: Basic Concepts, Theory and Applications. Springer, Boston, MA (2009)
11. Özdür, N.A., Üçel, I.B., Yang, J., Aydıner, C.C.: Residual intensity as a morphological identifier of twinning fields in microscopic image correlation. Exp. Mech. (2020). https://doi.org/10.1007/s11340-020-00672-8
12. Pan, B., Li, K., Tong, W.: Fast, robust and accurate digital image correlation calculation without redundant computations. Exp. Mech. **53**, 1277–1289 (2013). https://doi.org/10.1007/s11340-013-9717-6
13. Yang, J., Bhattacharya, K.: Augmented Lagrangian digital image correlation. Exp. Mech. **59**, 187–205 (2019). https://doi.org/10.1007/s11340-018-00457-0
14. Rueden, C.T., Schindelin, J., Hiner, M.C., et al.: ImageJ2: ImageJ for the next generation of scientific image data. BMC Bioinformatics. **18**, 529 (2017). https://doi.org/10.1186/s12859-017-1934-z
15. Schindelin, J., Arganda-Carreras, I., Frise, E., et al.: Fiji: an open-source platform for biological-image analysis. Nat. Methods. **9**, 676–682 (2012). https://doi.org/10.1038/nmeth.2019
16. Preibisch, S., Saalfeld, S., Tomancak, P.: Globally optimal stitching of tiled 3D microscopic image acquisitions. Bioinformatics. **25**, 1463–1465 (2009). https://doi.org/10.1093/bioinformatics/btp184
17. Pan, B.: Full-field strain measurement using a two-dimensional Savitzky-Golay digital differentiator in digital image correlation. Opt Eng. **46**, 033601 (2007). https://doi.org/10.1117/1.2714926
18. Oliphant, T.E.: Python for scientific computing. Comput. Sci. Eng. **9**, 10–20 (2007). https://doi.org/10.1109/MCSE.2007.58
19. Virtanen, P., Gommers, R., Oliphant, T.E., et al.: SciPy 1.0: fundamental algorithms for scientific computing in Python. Nat. Methods. **17**, 261–272 (2020). https://doi.org/10.1038/s41592-019-0686-2
20. Travis, O.: Guide to Numpy. Trelgol Publishing (2006)
21. van der Walt, S., Colbert, S.C., Varoquaux, G.: The NumPy array: a structure for efficient numerical computation. Comput. Sci. Eng. **13**, 22–30 (2011). https://doi.org/10.1109/MCSE.2011.37
22. Hunter, J.D.: Matplotlib: a 2D graphics environment. Comput. Sci. Eng. **9**, 99–104 (2007). https://doi.org/10.1109/MCSE.2007.55
23. Van Der Walt, S., Schönberger, J.L., Nunez-Iglesias, J., et al.: Scikit-image: image processing in python. PeerJ. **2014**, 1–18 (2014). https://doi.org/10.7717/peerj.453

Chapter 2
Time-Dependent Materials Response of Transverse Impact on Model Beams

Dulal Goldar

Abstract Author studied model beams (with equal overhang) simply supported beam like a bridge girder. The objective of this study is to understand TDM (Time Dependent Materials) response due to transverse impact experimentally.

Two types of experiments were carried out.

First dynamic photoelasticity was used to study transverse impact on urethane rubber beam (PSM-4) by free fall of a striker from a height of 176.4 mm employing Fastax framing camera (12,000 frames per second). Isochromatic fringe photographs were recorded in a light-field polariscope for central and non-central impact on simply supported beam with equal overhang for three different types of spans, namely, 90 mm, 120 mm and 150 mm with three different mass of strikers 10.52 g, 14.02 g and 17.53 g, respectively. Beam-striker weight ratio (2.675), height of fall, and dimensions of beam were kept constant.

From these experiments author finds beam span of 120 mm is critical from designer's point of view.

Author carried out second set of experiments with 120 mm beam span made of different materials to understand time dependent material response due to central impact.

A contact force transducer was fabricated (Goldar et al., Proc. 40th Anniversary Meeting, Spring Meeting, Society for Experimental Mechanics, Cleveland, Ohio, USA, pp 187–190, 1985) (with quartz crystals) to measure impact force using charge amplifier and storage oscilloscope. Second set of experiments were conducted on beams made of three different materials, namely, PMMA, Aluminium and layer composite PMMA-AL-PMMA with 120 mm beam span. The mass of striker was 41 g for central impact only.

Using electrical resistance strain gauges, contact force transducer peak-tensile strain vs. time and contact force vs. time were recorded to study time dependent material response.

To understand TDM response a nomogram indicating normalized Hertz's constant, striker-beam weight and peak-tensile strain for the beams made of different materials were plotted.

In addition peak-tensile strains vs. time in PMMA beam impacted centrally by different strikers were also recorded.

Another nomogram indicating normalized Hertz's constant, striker-beam weight ratio and peak-tensile strain for the PMMA beam under central impact were plotted.

Presence of small-amplitude 'precursors' and small-amplitude higher frequency oscillations in the strain-histories recorded and identified.

Key words Dynamic photoelasticity · Fastax framing camera · Urethane rubber beam · Contact force transducer · Electrical resistance strain gauge · beam · Transverse impact

2.1 Introduction

In modern times dynamic stress analysis is becoming increasingly important for various engineering applications. These applications may range from analysis of onshore and offshore structures subjected to earthquake, high velocity wind, wave – action and explosive blast loading to analysis of surface- and airborne-transport vehicles and other high-speed machinery; and last not the least to estimate stresses in biological system subjected to impulsive loading conditions. Collisions and impact by projectiles can set up dynamic stresses in all these situations. The different time histories of loading can produce different responses to the same system, due both to the variation in the magnitude of inertia forces so generated and also to that in material properties at different rates. Due to its importance from point of engineering design and also from point of a better

D. Goldar (✉)
Delhi College of Engineering (Now DTU), Delhi, India

A. Amirkhizi et al. (eds.), *Challenges in Mechanics of Time Dependent Materials, Mechanics of Biological Systems and Materials & Micro-and Nanomechanics, Volume 2*, Conference Proceedings of the Society for Experimental Mechanics Series,
https://doi.org/10.1007/978-3-030-86737-9_2

understanding of material[1] properties, the subject of dynamic stress analysis and material engineering have attracted the attention of many research workers in the past. Although significant amount of insight in these respects have been developed, the subject being rather intricate would continue to attract attention of more research workers in future.

2.1.1 *Experimental Setup*

Transverse Impact load was induced with the help of a freely falling striker on a urethane rubber (PSM-4) beam model (size: 253 mm long × 24.3 mm deep × 12.3 mm thick) for photoelastic studies [1]. Different overhang ratios, defined as length of overhang/total length of the beam were considered for three different spans, namely 90 mm, 120 mm and 150 mm. For the present study, ratio of weight beam between supports and weight of striker was kept constant (2.675).

System for Generating Impact

The system for generating impact was essentially consists of an electromagnet (60 V DC, 40 mA), an aluminium guide pipe (200 mm long, 16 mm dia.) and three different strikers for three cases. The assembly of electromagnet (Fig. 2.1) was fixed centrally over ab aluminium plate (540 mm × 302 mm × 3.4 mm thick). Below this electromagnet the aluminium guide pipe was fitted vertically with the help a sleeve of the same material. This sleeve was connected with the bottom of the horizontal plate with three brass screws. At the end of guide pipe, vertical cuts of 30 mm height on diametrically opposite sides were made for recording movements of striker on photographic film negative. The horizontal plate was supported on four vertical mild steel rods of 12.5 mm dia. and 800 mm height. Horizontal position of the aluminium plate was adjusted with the help of two sets of mild steel rings and clamping screws on each vertical rod. The horizontality of the supporting aluminium plate was checked with help of spirit level. The transverse impact load was induced with the help of a freely falling striker on a urethane rubber (PSM-4) beam model (size: 253 mm long × 24.3 mm deep × 12.3 mm thick) for photoelastic studies [1]. Different overhang ratios, defined as length of overhang/total length of the beam were considered for three different spans, namely 90 mm, 120 mm and 150 mm. For the present study, ratio of weight beam between supports and weight of striker was kept constant (2.675).

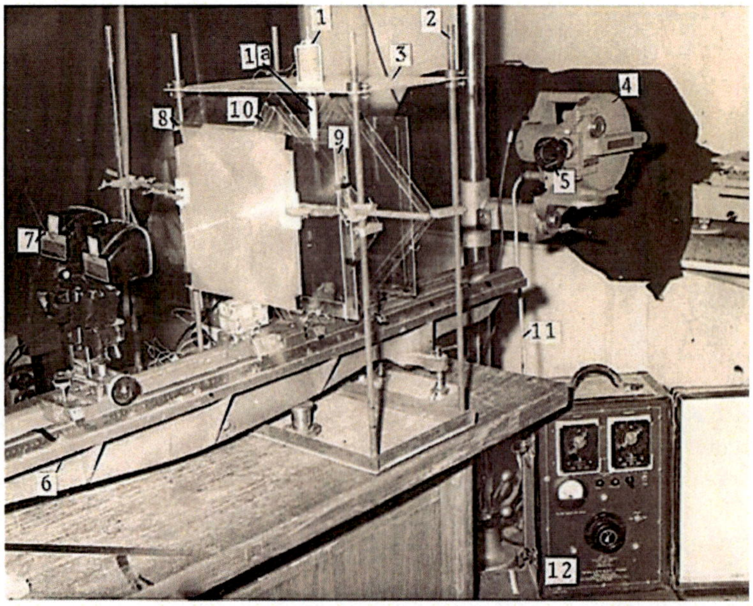

Fig. 2.1 Photograph of experimental setup for dynamic photoelastic studies. (1) Electromagnet (60 V DC, 40 mV), 1a. Guide pipe, (2) Vertical mild steel rod for supporting electromagnet, (3) Aluminium plate, (4) Fastax, 16 mm framing camera, (5) Monochromator, (6) Optical bench, (7) Light source (Sun Gun-II, 800 W), (8) Fresnel lens and diffuser plate, (9) Plane polaroid, (10) Quarter-wave plate, (11) Signal from time-marker, (12) Goose control unit

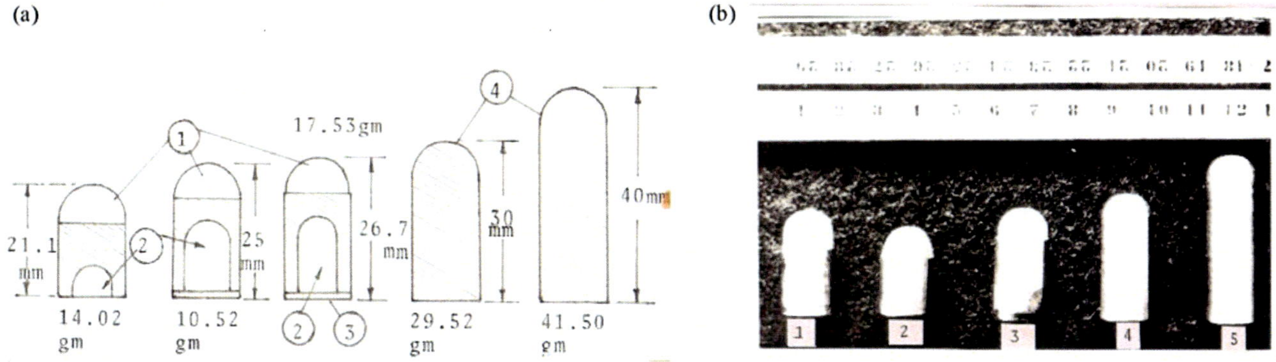

Fig. 2.2 Different strikers. (**a**) (1) Araldite hemispherical tip, (2) Cavity for adjusting weight of strikers, (**3**) Mild steel disc for holding properly with electromagnet, (4) Mild steel strikers. (**b**) (1, 2, 3) Strikers with araldite hemispherical tip of weight 10.52 g, 14.02 g and 17.53 g, respectively. (4, 5) Mild steel strikers of weight 29.52 g, 41.50 g, respectively

System for Generating Impact

The system for generating impact was essentially consists of an electromagnet (60 V DC, 40 mA), an aluminium guide pipe (200 mm long, 16 mm dia.) and three different strikers for three cases. The assembly of electromagnet (Fig. 2.1) was fixed centrally over an aluminium plate (540 mm × 302 mm × 3.4 mm thick). Below this electromagnet the aluminium guide pipe was fitted vertically with the help a sleeve of the same material. This sleeve was connected with the bottom of the horizontal plate with three brass screws. At the end of guide pipe, vertical cuts of 30 mm height on diametrically opposite sides were made for recording movements of striker on photographic film negative. The horizontal plate was supported on four vertical mild steel rods of 12.5 mm dia. and 800 mm height. Horizontal position of the aluminium plate was adjusted with the help of two sets of mild steel rings and clamping screws on each vertical rod. The horizontality of the supporting aluminium plate was checked with help of spirit level.

Details of Strikers

Three different strikers (Fig. 2.2) weighing 10.53 g, 14.02 g and 17.53 g were made of mild steel with a hemispherical tip of araldite. It was intended to produce transverse impact between two nearly similar materials and not widely dissimilar materials such as urethane rubber and mild steel. Therefore, a hemispherical piece of araldite (Ciba CY230 with hardener HY951) was cemented to the mild steel portion on the impacting side of all the three strikers. It may be seen in Fig. 2.2a, b that the above strikers had machined cavities which were provided to accurately adjust self-weights for maintaining beam/striker weight ratio constant (2.675).

Arrangement of Support for Beam

The beam with equal overhang on either side was simply supported over two mild steel wedges (Fig. 2.3). With the aid of slots and clamping screws, the distance between the wedges could be continuously adjusted between 80 mm and 290 mm. this supporting system was clamped on another fixture (in which coarse and fine adjustments were provided for both horizontal and vertical movements) available with optical bench (Fig. 2.1). For each simply supported beam span with equal overhang (i.e. 90 mm, 120 mm and 150 mm) marking with pen-ink were made for central point on top surface of urethane beam and two support points on bottom of beam. Similar markings were also made for the study of quarter-span (non-central) impact points. The horizontality of beam-wedge support and verticality of pipe were checked thoroughly.

After these adjustments the model beam was placed as per markings over the wedge support for central as well as non-central cases. For all the three cases for central as well as non-central impact cases and for different strikers the height of fall was kept same, namely 176.4 mm. Accordingly clear gap between top surface of urethane model beam and bottom of aluminium guide pipe marginally adjusted for different strikers employed.

The strikers were kept at the above height by energizing the electromagnet through a DC power supply. A low supply current was maintained to minimize the time delay for de-energizing the electromagnet.

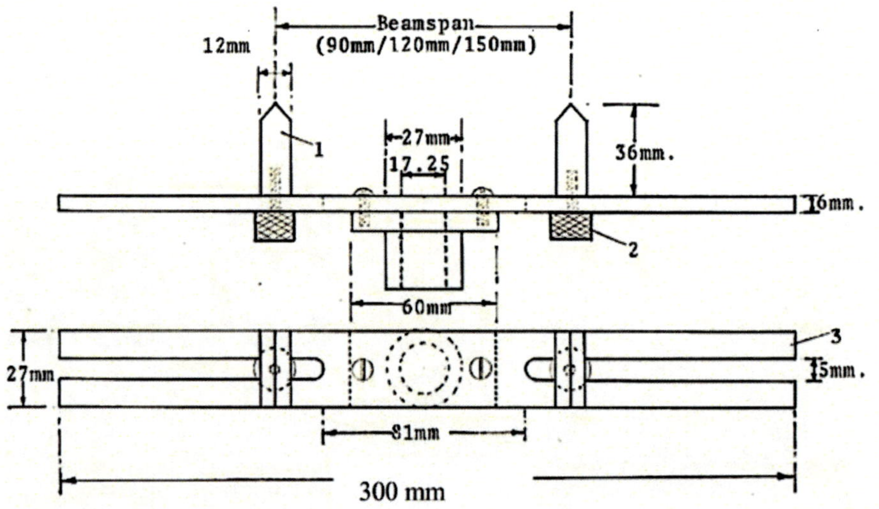

Fig. 2.3 Arrangement of support for beam. (1) Wedge, (2) Clamping screw, (3) Beam support base

Arrangement for Fringe Photography

For photoelastic study a diffused light polariscope was designed. Two sets of plane polaroid and quarter-wave plate (300 mm × 300 mm size) were mounted on square frame made of 3 mm thick plexi glass (PMMA) sheet. The elements were placed in such a way that polariscope may be set in either dark- or light-field arrangements. For illumination of the urethane rubber beam, two Sun Gun-II, 900 W were placed side by side and clamped on a fixture of the optical bench. The illumination was made uniform by a combination of a diffuser plate and a Fresnel lens. An interference filter of band pass less than 100 Å (10^{-8} m) Model-068, Photoelastic Inc.,USA was used before the Fastax camera lens to render the light monochromatic.

For the present study the optical elements were arranged in such a way as to produce light-field background. This arrangement helped for recording clearly the supports and movement of the falling striker through the slot cut in the guide pipe and also through the gap between the top face of the beam and the guide pipe. Again from this record contact velocity of the striker was determined. Also the clearance between the guide pipe and beam helped to reduce the resistance to the motion of the striker.

The Fastax Camera

Wollensak Fastax (16 mm) framing camera [2] was used for fringe photography. The camera was operated with a typical framing speed of 12,000 f.p.s. (frames per second). For the first trial 120 ft (36.58 m) length of film used. For subsequent records, however, 60 ft (18.29 m) length of film was adequate. The object lens of the camera had a focal length of 2.0 in. (50.8 mm), and a maximum aperture of f/2.0. The image of the unloaded beam was first focussed by the parallax method with the help of telescopic-view-finder, and subsequently checked on the film-plane by statically loading the urethane beam model so as to develop a typical fringe pattern.

Method of Synchronization

A schematic diagram showing the method of synchronization [1, 3] used in fringe photography is indicated in Fig. 2.4. Before recording a particular fringe photograph, the following operations were conducted in a sequence:

1. The time of fall of the striker was measured by 'starting' a timer circuit by operation of the electromagnet operating switch and 'stopping' the timing circuit by a metal foil short-circuiting. The time of fall was used in setting 'Event marking' time of the 'Goose-Control Unit' (Fig. 2.4).
2. Depending upon the framing speed desired and the total length of the film used, the 'Voltage Control' and 'Camera time' of the 'Goose Control Unit' were then set.

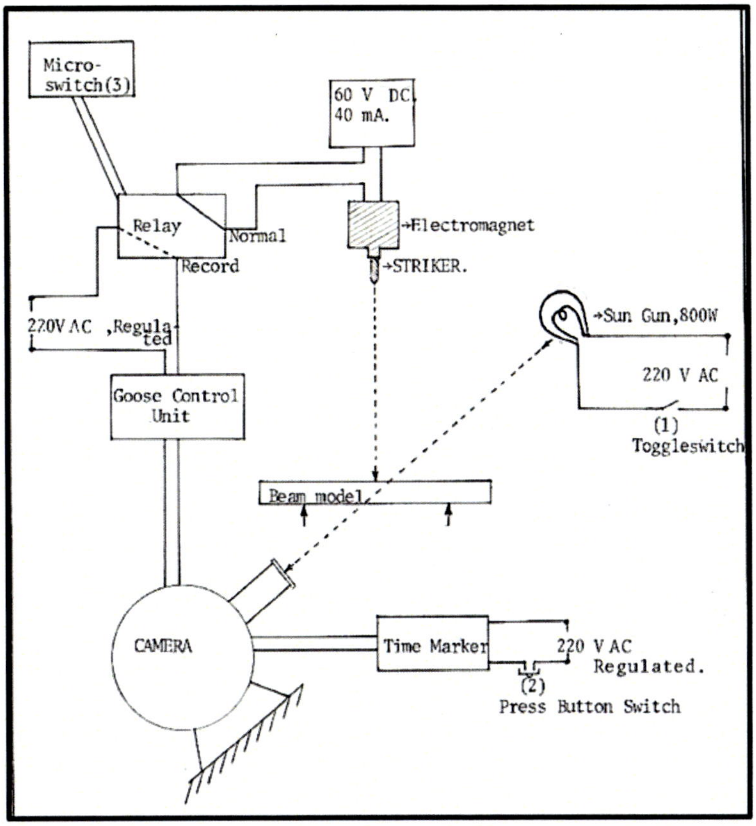

Fig. 2.4 Method of synchronization

3. The Relay (Fig. 2.4) was at 'Normal' position by which the electromagnet was energized.
4. All the laboratory lights were switched off and the Sun-gun toggle switch was switch on.
5. The 'Press-button' switch of the 'Time-marker' was pressed.
6. The 'Micro-switch' (Fig. 2.4) was operated which set the Relay in the 'Record' position. The Relay de-energized the electromagnet and simultaneously switching on the 'Goose-Control unit'. The 'Goose-Control unit' in turn started motors of the framing camera. By the time the striker hit the beam model, the desired camera speed was achieved. The camera was automatically switched off by the 'Camera Time' control provided in the 'Goose-Control unit', and therefore, the recording was completed.
7. The 'Time-marker' press-button switch and Sun-gun toggle switch were then switched off.

Photographic Film

KODAK 2498 RAR (250 ASA) 16 mm roll films were used for fringe photography. KODAK developer, freshly prepared from chemicals, was used for developing these film rolls.

Enlarged prints, both continuous and discrete (selected frames), were prepared from film negatives to study the history of fringe formation and to identify precisely fringe orders (Figs. 2.5, 2.6, 2.7 and 2.8). Only photographs showing continuous prints from frame number 0 to 17 are shown for 150 mm, 120 mm and 90 mm beam span for the sake of brevity for central as well as quarter-span impact cases. Enlargement used for selected frames was significantly greater than that used for continuous printing. Some such selected frames (Figs. 2.6 and 2.8 for central and quarter-span impact cases) were printed compositely along with a transparent millimetre grating so as to facilitate the plotting of free-boundary stress distributions (Figs. 2.9 and 2.10 for central and quarter-span impact cases for specific instances of impact) in the beam.

From isochromatic fringes photographs [3–5] stress wave propagation, [6, 7] contact velocity [8] boundary stress and deflection [9] were recorded for both central and non-central impact cases. Utilizing the results it was observed that 120 mm beam case [3] to be important from designer's point of view.

Fig. 2.5 Isochromatic fringe pattern for central impact loading

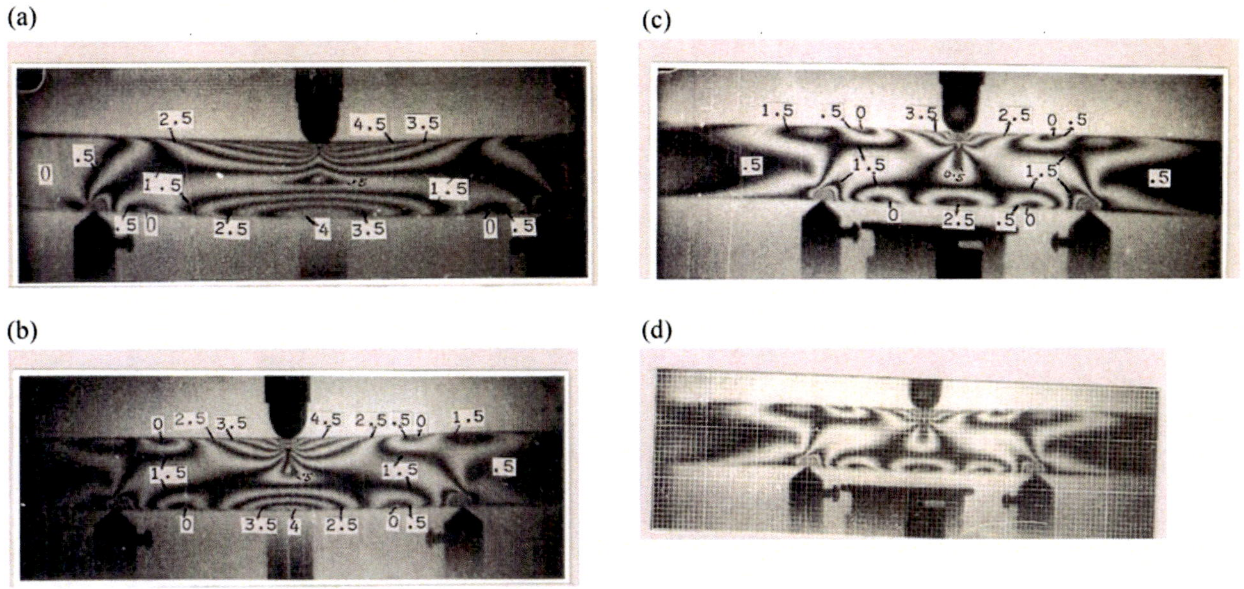

Fig. 2.6 Enlarged photographs. (**a**) 150 mm Beam Span, Frame No. 51, Time = 4936 μs. (**b**) 120 mm Beam Span, Frame No. 30, Time = 2523 μs. (**c**) 90 mm Beam Span, Frame No. 20, Time = 1897 μs. (**d**) 90 mm Beam Span, Frame No. 20, Time = 1897 μs. with mm-grating

Discussion of Results for Dynamic Photoelastic Studies

Paldas [9] reported the mechanism of isochromatic fringe formation in simply supported urethane rubber beams (Hysol 8705) subjected to central impact loading. The collision velocity was 3.02 m/s, and the span-depth ratio and beam-striker weight ratio ranged between 3.97 and 5.87 and between 1.614 and 8.168, respectively. In two of the four cases reported, there occurred two sub-impacts and the remaining two cases single-sub-impact occurred. A systematic study of fringe photographs revealed that the mechanism of isochromatic fringe formation during the early stages of impact was identified for all the cases, namely generation of 'semi-infinite plate' behaviour, reflection of stress-waves from bottom-fibre of the beam, and then generation of 'elastically supported' beam behaviour. The mechanism of fringe formation during the later stages of impact was different for four cases, depending upon whether a second-sub-impact between the beam and the striker occurred or not.

In the present study [1, 3] a urethane rubber (PSM-4) beam model was subjected to both central- and quarter-span-impact loadings were simply supported with considerable amount of overhang (overhang ratio ranged between 0.204 and 0.322), and were impacted with a constant velocity ranging between 1.163 and 1.700 m/s. The span-depth ratios varied between 3.711 and 6.186, while the beam-striker weight ratio was kept constant (2.675). In the present situation a single sub-impact occurred always (Fig. 2.5). This phenomenon is attributed to the inertia overhang portions of the beam. Comparing the fringe pattern obtained in the present study with that of Paldas [9], it may be stated that the mechanism of isochromatic fringe formation during the early stages of impact remains unchanged irrespective of the generation of 'semi-infinite plate' behaviour, reflection of stress waves from the bottom-fibres of the beam and generation of 'elastically supported' beam behaviour. This behaviour is expected, since the stress waves would require certain amount of time to travel the distance from the point of impact to the support locations and further onwards, and till such time the support reactions are adequately mobilised, there cannot be any difference in the behaviour of the beam model.

For the situations of quarter-span impact loadings an additional feature could be recognised during the early stages of impact, namely the 'simple compression' behaviour. For all the three cases for spans 90 mm, 120 mm and 150 mm studied the quarter-span impact loading took place at the left-quarter-span location, consequently the point of loading had proximity to the left-hand support locations. A direct 'fringe-link' between the point of impact and the left support proved the existence of such 'simple compression' behaviour. For 90 mm beam span, such a 'simple compression' behaviour was most pronounced as compared to 120 mm and 150 mm beam spans. From isochromatic fringe photographs (Fig. 2.7) such 'simple compression' behaviour was longest, i.e. 7.34 ms for 90 mm beam span and was shortest, i.e. 2.11 ms for 150 mm beam span. For the 120 mm beam span the duration of such a 'simple compression' behaviour was 2.75 ms close to the 150 mm beam span and was quite far removed from that of 90 mm beam span. Thus 90 mm beam case was a special situation.

From enlarged isochromatic photograph as show in Figs. 2.5 and 2.6 for central and Figs. 2.7 and 2.8 for quarter-span impact loading a 'zero-order' fringe appeared on the top-fibre of the beam during the early stages of impact load, and the

Fig. 2.7 Isochromatic fringe pattern for quarter—span impact loading

'elastically supported' beam behaviour could be identified from the appearance of such zero-order fringes. Since a zero-order fringe at a free-boundary indicates stress-free state, the location of zero-order fringe on the top-boundary has been indicated as location of stress-free point on the top-boundary. From this speed of travel, stress-free point was determined and reported elsewhere [1, 7].

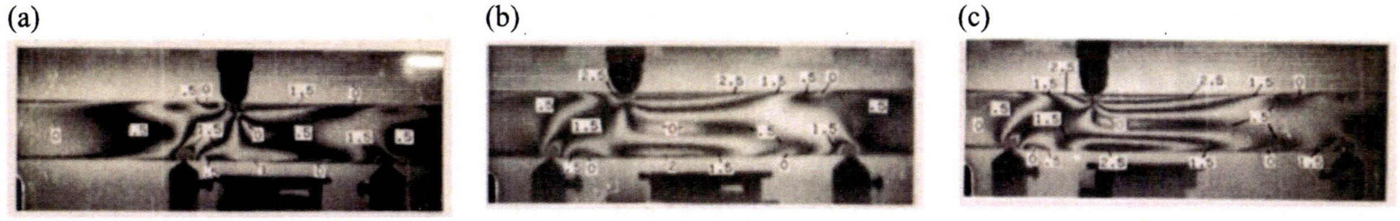

Fig. 2.8 Enlarged photograph of selected frames. (**a**) 90 mm beam span, Frame No. 13, Time = 1352 μs. (**b**) 120 mm Beam Span, Frame No. 40, Time = 4238 μs. (**c**) 150 mm Beam Span, Frame No. 35, Time = 3361 μs

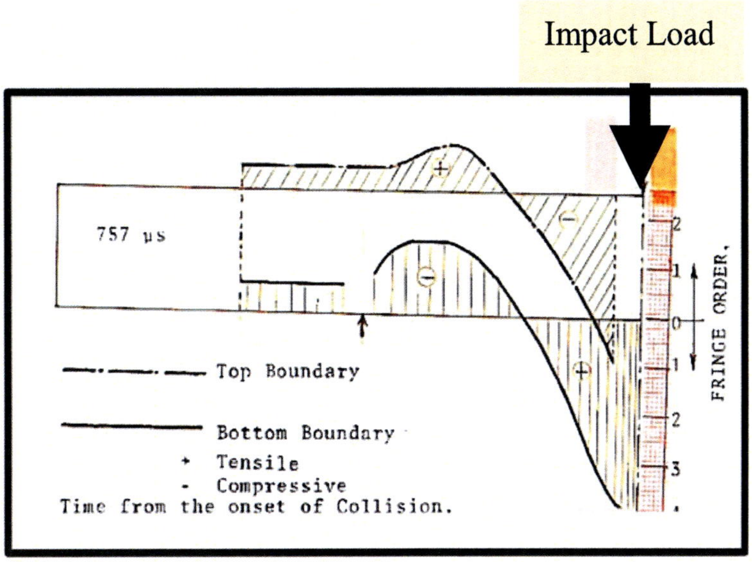

Fig. 2.9 Free-boundary stress distribution for 120 mm beam-span under central impact loading

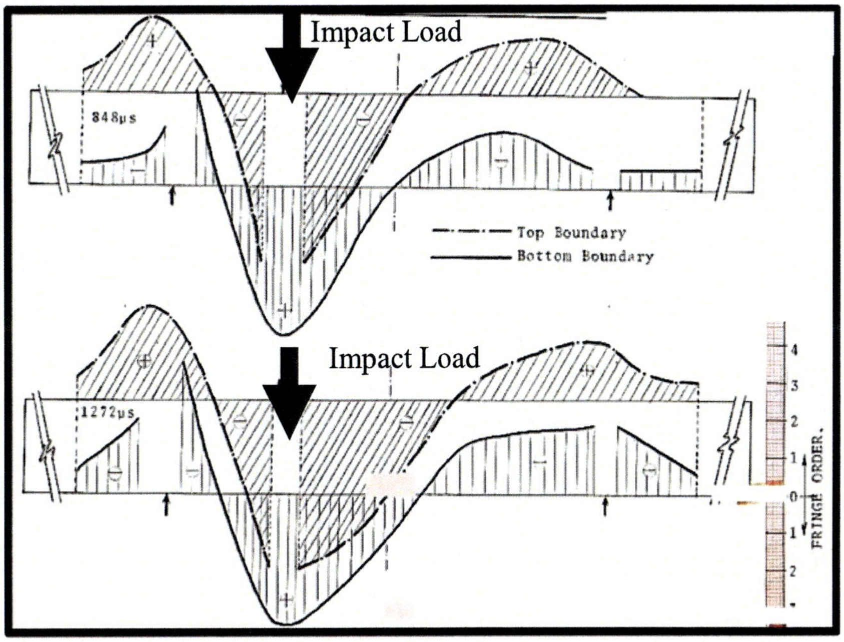

Fig. 2.10 Free-boundary stress distribution for 120 mm beam-span under quarter-span impact loading

2.2 Experimental Study of Transverse Impact on Aluminium, PMMA and PMMA-Al-PMMA

Next, beams made of two other materials, one metallic (Aluminium) and other polymeric (PMMA) and composite PMMA-Aluminium-PMMA, having identical supports and geometric conditions as the model photoelastic beams, were subjected to central impact loadings for 120 mm span simply supported. The weight of the striker, however, was increased to 41.5 g. The experimental set up is shown in Fig. 2.11. The details are reported elsewhere [1, 7]. The contact force- and strain-histories in these beams were also found out by using a force transducers and electrical stain gauges respectively. Utilizing these data, an attempt was made to correlate the contact force, the maximum tensile stress and a material constant for such simply supported beams with overhang subjected to low velocity impact (<2 m/s) by a light striker.

2.2.1 Peak-Tensile Strains in Perspex, Perspex-Aluminium-Perspex and Aluminium Beams Under Central Impact Loading

Peak-tensile strain for the above three beams were experimentally obtained and presented in Table 2.1 along with Hertz's constants between the striker's tip and beam materials.

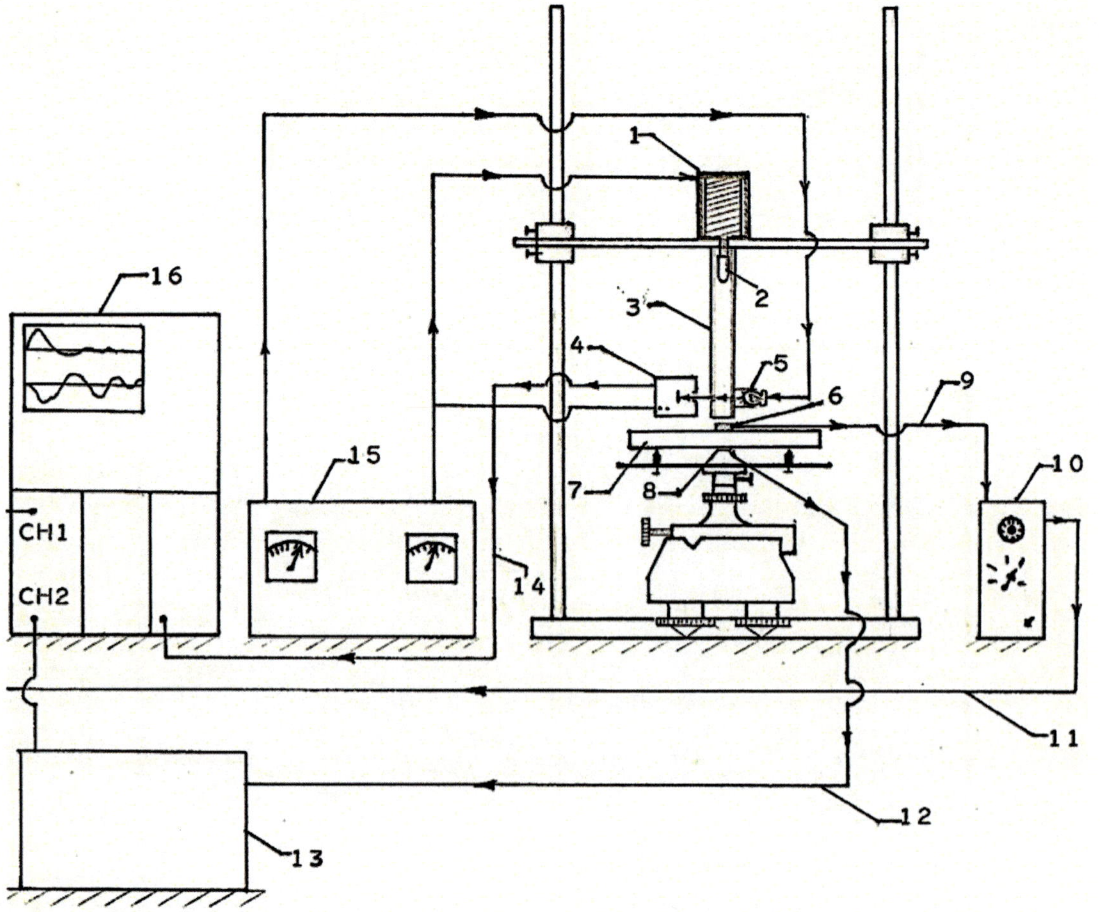

Fig. 2.11 Experimental arrangement for measurement of contact-force, strain and contact velocity. (1) Electromagnet (60 V DC, 40 mA), (2) Striker, (3) Guide Pipe, (4) Photo-transistor, (5) Light Source (10 V DC), (6) Contact-Force Transducer, (7) Beam Model, (8) Strain Gauge, (9) Signal from Force-Transducer, (10) Charge Amplifier (KISTLER), (11) Signal from Charge Amplifier, (12) Signal from Strain Gauge, (13) Bridge Balancing and Amplifying Unit, (14) Signal from Photo-transistor to External Triggering of Oscilloscope, (15) Variable DC Power Supply, (16) Storage Oscilloscope

Table 2.1 Peak-tensile strains in different beam materials under central impact loadings

Beam material	Hertz's constant ($cm/g^{2/3}$)	Hertz's constant normalized for impact on Perspex beam	Striker weight/beam weight	Peak-tensile strain obtained from experiments (μ)
Perspex	7.42×10^{-6}	1.00	0.93	853
Perspex-AL-Perspex	7.42×10^{-6}	1.00	0.65	747
Aluminium	1.42×10^{-6}	0.10	0.41	128

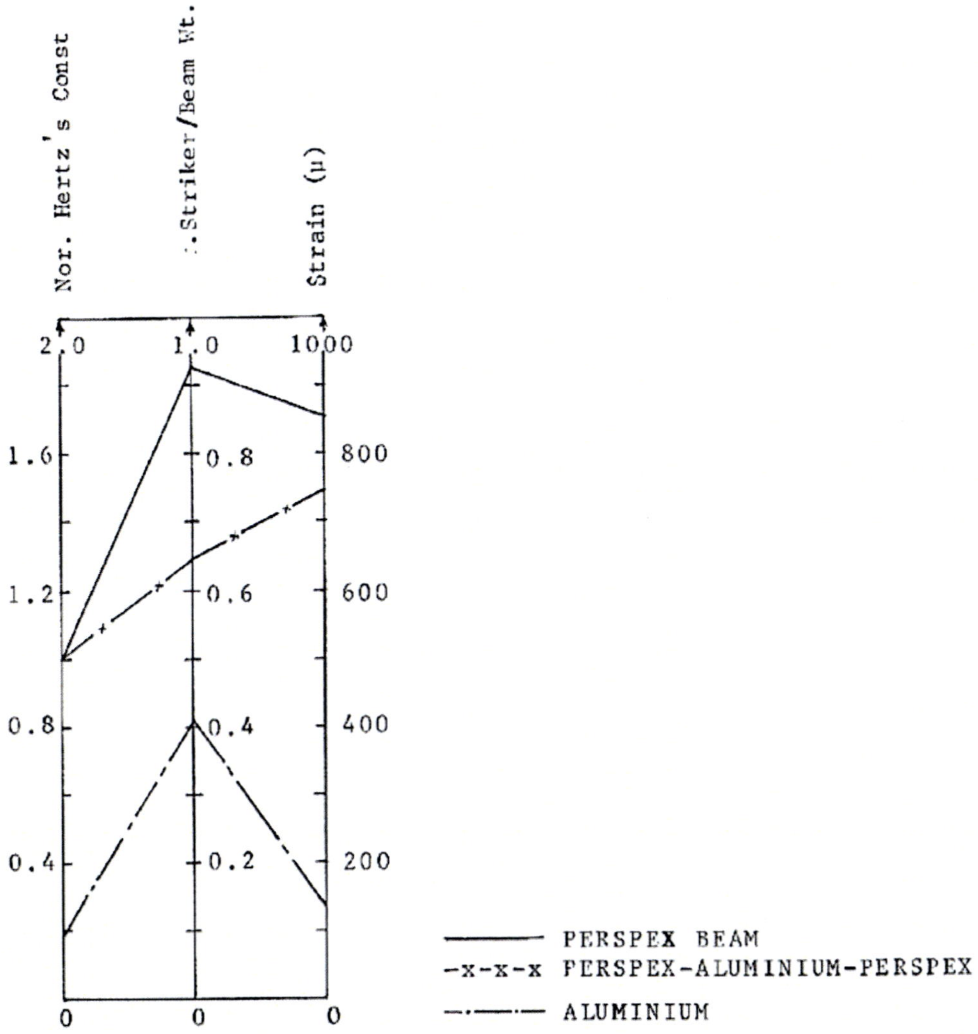

Fig. 2.12 Nomogram indicating normalized Hertz's constant, Striker-beam weight ratio and Peak-tensile strain for the beam of different materials

The values of Hertz's constant have been normalized in respect of the Perspex beam and the striker/beam weight has been calculated. The values of normalized Hertz's constant, striker-beam weight ratio and the peak-tensile strains have been plotted in a nomogram and the same is presented in Fig. 2.12.

2.2.2 *Peak-Tensile Strains in the Perspex Beam Under Central Impact by Different Strikers*

A composite plot of strain-histories for Perspex beam impacted by three strikers namely, 14.02 g, 29.52 g and 41.50 g of weight are shown in Fig. 2.15. The Hertz's constant in respect of araldite and Perspex (corresponding to striker weight 14.02 g was calculated [1]. In Table 2.2 striker weight, Hertz's constant, striker weight/beam weight and peak-tensile strain are presented.

Table 2.2 Peak-tensile Strains in Perspex beam under central impact loadings

Striker weight (g)	Hertz's constant (cm/g$^{2/3}$)	Hertz's constant normalized in respect of striker of 41.5 g weight	Striker weight/beam weight	Peak-tensile strain (μ)
41.50	7.42×10^{-6}	1.00	0.930	947
29.52	7.42×10^{-6}	1.00	0.66	815
14.02	9.55×10^{-6}	1.29	0.313	361

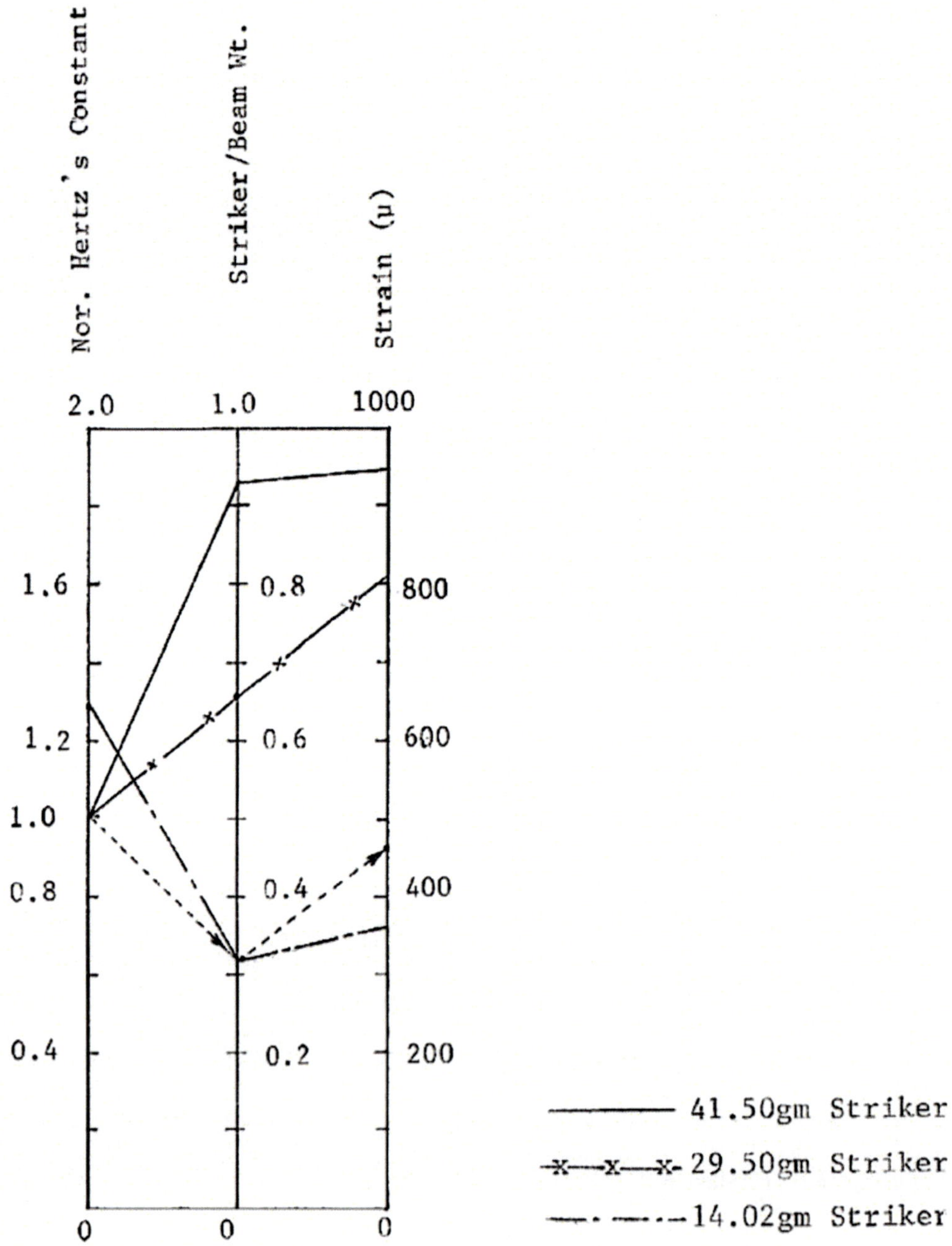

Fig. 2.13. Nomogram indicating normalized Hertz's constant, Striker-beam weight ratio and Peak-tensile strain for the Perspex-beam under Central impact loading with different Strikers

The values of Hertz's constants have been normalized in respect of striker weight, 41.50 g, and the striker-beam weight ratios have been calculated. These values of normalized Hertz's constant, striker-beam weight ratio and peak-tensile strains for the Perspex beam subjected to central impact loadings have been plotted in a nomogram and is shown in Fig. 2.13 for the

14.02 g striker, the peak-tensile strain for a 'normalized Hertz's constant' of 1.00 has been estimated from a reciprocal relationship between 'normalized Hertz's constant' and peak strain, and the same is also shown in nomogram (Fig. 2.13) with dotted line.

2.3 Conclusion

2.3.1 *Presence of Small-Amplitude 'Precursor' in the Strain-Histories Record*

Goldsmith et al. [10] reported the presence of 'precursor' in the strain-history recorded with the help of strain-gauges. In the present work also a number of strain-histories were recorded with the help of strain-gauges on the top and bottom boundaries of the beam (with the exception of urethane rubber beam), and are presented in Figs. 2.14 and 2.15. It may be seen from these

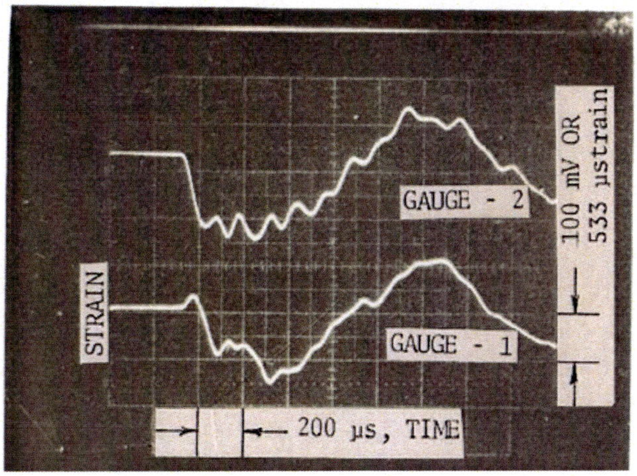

Fig. 2.14. Oscilloscope traces showing strain-histories for Perspex-beam under central impact loading by 41.5 g striker (mild steel)

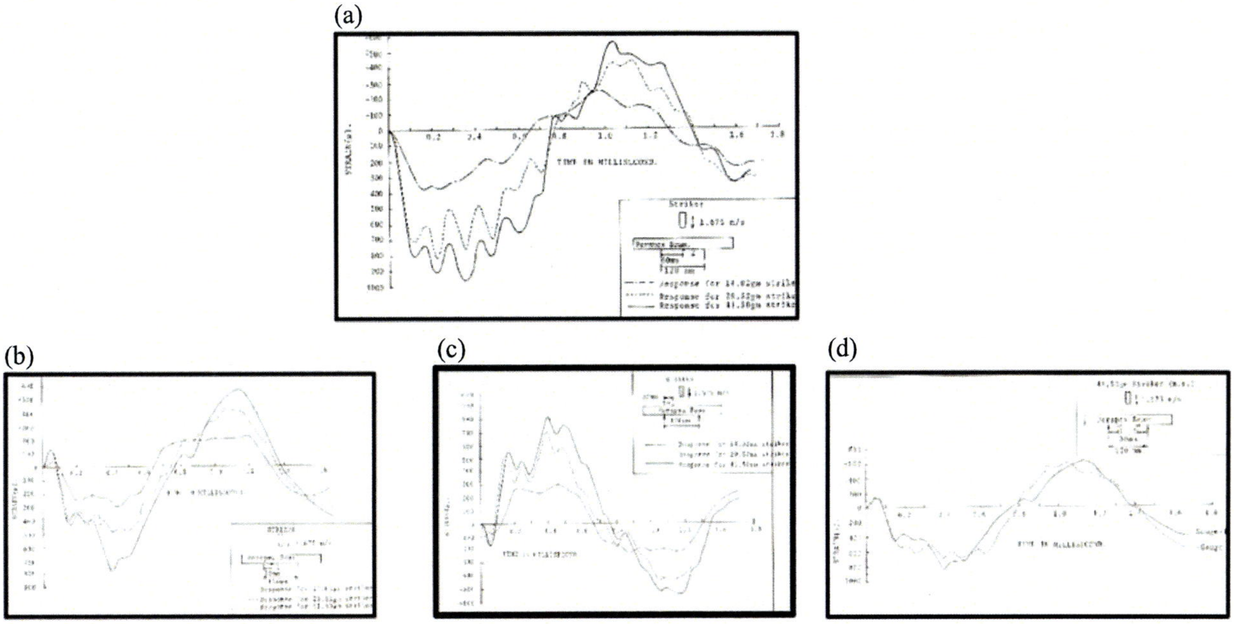

Fig. 2.15 Composite plot of strain-histories at (**a**) central-span location (**b**) left- quarter span location bottom fiber (**c**) left-quarter span location top fiber for Perspex-beam for 14.02 g, 29.52 g and 41.50 g strikers and (**d**) left-quarter and right-quarter span location for 41.50 g striker

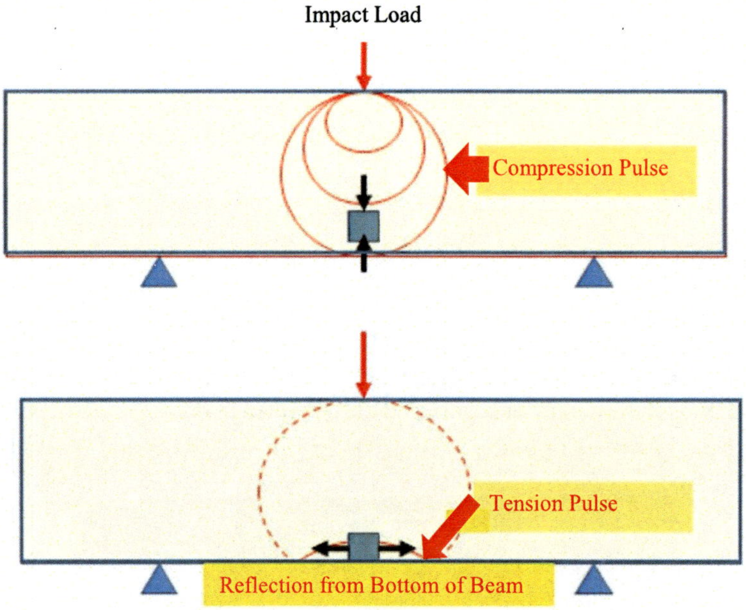

Fig. 2.16 Stress-waves propagation

figures that a small-amplitude 'precursor' was generally present in these strain-histories both at the top and bottom boundaries, especially at quarter-span locations. Again, for the urethane rubber beam model, free-boundary stress distributions at selected instants of time were drawn from the fringe photographs and are presented in in Figs. 2.9 and 2.10. It may also be seen from these figures that a similar 'precursor' in the stress-history is also generally present. Paldas [9] reported that at an early stage of impact loading on a simply supported beam (without overhang), the fringe pattern developed resembled those produced in a 'semi-infinite plate' subjected to a line load on the straight edge, producing compressive stresses at the lower edge of the central section. Then the stress waves were reflected from the lower edge of the beam and subsequently developed tensile stresses. The 'precursor' recorded in the present study relate to generation of such an opposite state of stress at early stage of impact in that prevailing later on at free-boundary point. In fact it may be seen from Figs. 2.9 and 2.10 that the magnitude of these 'precursor' along the top boundary becomes significant beyond the quarter-span locations and reached a peak-value over the supports. It implies, therefore, that the material employed in preparing these beams should be capable of withstanding significant amount of tensile stresses the top fibres also, especially in the vicinity of support locations (Fig. 2.16).

Acknowledgements The author is thankful to the authorities of Terminal Ballistic Research Laboratory, Chandigarh, India for providing all necessary facilities.

References

1. Goldar, D.: Photoelastic Studies of Transversely Impacted Simply Supported Beams. Ph.D Thesis by, August 1981, Panjab University
2. Operators Manual, Fastax High Speed Cameras, Category-I through VI, Wollensak Division, Co. Rochester, NY, USA
3. Goldar, D., Verma, S.P., Paldas, M.: Photoelastic studies of transversely impacted beams with over-hang. In: Proc. Symposium on 'War -Head Technology' Organized by Terminal Ballistic research Laboratory (TBRL), Chandigarh, March 1983, presented the paper
4. Goldar, D.: Dynamic response of transversely impacted simply supported beams with equal overhang. In: Proc. VII-Intel. Congress on Experimental Mechanics, Las Vegas, Nevada, USA
5. Goldar, D.: Full-Field Dynamic Photoelastic Studies of Transversely Impacted Beams. Paper#OS1W0028, ATEM2003, JSME-MMD, Nagoya, Japan (2003)
6. Kolsky, H.: Stress Waves in Solids, p. 84. Dover Publication Inc. (1963)
7. Goldar, D.: Elastic wave propagation in transversely impacted beams. In: Keynote Speaker, 8th. ISEM, Sendai, Japan November 03–06, 2013 (2013)
8. Goldar, D.: Experimental determination of contact velocity (Paper code number A147). In: 11th Asian Conference on Experimental Mechanics, 2012 Society for Experimental Mechanics Fall Conference and 7th International Symposium in Advanced Science and Technology in Experimental Mechanics. The conference will take place in The Grand Hotel, Taipei, Taiwan, on November 8–11, 2012 2012

9. Paldas, M.: Deflection and stresses in simply supported beams under impact loading. J. Inst. Eng. **54**(6), 219–224 (1974)
10. Goldsmith, W., Norris Jr., G.W.: Stresses in curved beams due to transverse impact. In: Proceedings U.S. Natl. Congr. Applied Mechanics, pp. 153–162. ASME (1958)
11. Goldar, D., Sethi, V.S., Khurana, O.P., Verma, S.R.: Development and calibration of a dynamic contact-force transducer. In: Proc. 40th Anniversary Meeting, Spring Meeting, Society for Experimental Mechanics, Cleveland, Ohio, USA, presented the paper and subsequently published in J. Experimental Mechanics, pp. 187–190 (1985)
12. Kuske, A.: Photoelastic stress analysis of machines under dynamic load. Exp. Mech. **17**(3), 88–96 (1977)

Chapter 3
Wearable Device for Tremor Suppression

Samuel E. Winston, Riley C. Dehmer, Joseph Horen, and Timothy A. Doughty

Abstract Parkinson's disease is a neurodegenerative disease that affects nearly a million people in the United States. Currently there is no cure for the disease, but there are many attempts to manage the symptoms. One effort uses assistive devices, which can help patients cope with the most common symptom: hand tremors. The goal was to design a noninvasive, adjustable device that effectively suppresses hand tremors. To begin, theoretical models were developed to gain an understanding of the governing principles involved in the hand–device interface. Simulated models gave insight to design considerations. Research moved into prototyping which involved sketching, modeling, and fabrication. Alongside prototyping, initial testing was performed to view qualitative tremor suppression. From this iterative process, two designs (The String and The Pin) met the essential criteria. These devices went through further testing at various adjustment settings to quantify their tremor suppression capabilities. The desired capabilities involved effectively suppressing frequencies from 4 to 12 Hz, since Parkinsonian tremors range from 4 to 6 Hz (Zach et al., J Parkinsons Dis 5:471–474, 2015) and essential tremors occur at higher frequencies. The String design had an adjustable effective range of 4.7–11.5 Hz whereas The Pin was only adjustable to be effective from 3.2 to 6.2 Hz. The effectiveness was defined as a reduction in tremor response by an order of magnitude. Additionally, once fixed to address a certain tremor frequency, The String design was able to effectively suppress a wider range of frequencies when compared to The Pin design (U + 00B1 2.3 Hz and U + 00B1 0.55 Hz).

Key words Frequency · Parkinsonian tremor · Essential tremor · Tremor suppression · Mechanical device · Assistive technology

3.1 Background

In 2010, a total of 680,000 people in the United States were affected by Parkinson's Disease (PD). That number is projected to grow to nearly a million people by 2020 [2]. PD is a movement disorder that worsens over time by damaging the nervous system. PD occurs when neurons in the brain become impaired or die. Many regions in the brain are affected, although the most common symptoms result from the loss of neurons in the *substantia nigra.* One of the primary symptoms of this disease is tremors: often in the hand, an involuntary, rhythmic back and forth motion that can involve the thumb and forefinger to appear as "pill rolling" [3].

The top three challenges PD individuals face are with nutrition, sleeping, and mental health. Nutrition is primarily impacted by the interaction between the tremors and the physical act of eating. Patients are easily discouraged from eating as they struggle to put food up to their mouth. Other symptoms, such as difficulty swallowing or muscle cramps, can enhance the struggle with maintaining a healthy diet [4]. Getting a healthy amount of sleep with PD is another challenge because tremors often keep patients awake, muscle cramps can cause sudden waking, and the treatment medication disrupts sleep. These difficulties are compounded by the role mental health plays in maintaining healthy sleep habits [5]. Mental health issues are extremely common among patients. It is estimated roughly half of all patients will experience anxiety and/or depression while combatting the disease. This increased mental strain is directly caused by how PD changes brain chemistry. On top of coping with poor mental health, it can even reduce how effective a treatment is for the individual [6].

Currently there is no cure for PD, however there are three main treatments to help with its symptoms: drug therapy, surgery, and lifestyle changes (i.e., diet and exercise). Drug therapy medications work by increasing dopamine in the brain, affecting

S. E. Winston · R. C. Dehmer · J. Horen · T. A. Doughty (✉)
Shiley School of Engineering, University of Portland, Portland, OR, USA
e-mail: winston21@up.edu; dehmer22@up.edu; horen19@up.edu; doughty@up.edu

A. Amirkhizi et al. (eds.), *Challenges in Mechanics of Time Dependent Materials, Mechanics of Biological Systems and Materials & Micro-and Nanomechanics, Volume 2*, Conference Proceedings of the Society for Experimental Mechanics Series,
https://doi.org/10.1007/978-3-030-86737-9_3

other neurotransmitters in the body, or trying to control non-motor symptoms. Symptoms improve at first with these medications but reappear over time as the drugs become less effective [3]. Patients are also required to strictly follow a time regimen for these medications, for instance some patients treated with carbidopa/levodopa require a dose every 2 h. Delaying the medications by more than an hour can cause immediate increases to the symptoms [7]. These drugs also have many side effects. Initial side effects include nausea, low blood pressure, restlessness, and drowsiness. While long-term side effects lead to hallucinations, psychosis, and dyskinesias (involuntary movements during time of peak benefit from dose of levodopa) [3]. Another treatment to aid with "off" time (periods where medications are not offering benefit) and dyskinesias is Deep Brain Stimulation. This is a surgical procedure that involves implanting electrodes into certain regions of the brain [8]. Although this improves motor control, it is very invasive and comes with a lot of risk to the patient.

Non-invasive assistive devices provide an attractive method of tremor suppression. The most affordable of which is a weighted utensil, however many studies have shown these utensils do not significantly reduce the effect of tremors. Some devices actively attempt to compensate for tremors by sensing the hand's motion and moving the end of a utensil. The cost of the device is rather expensive, a starter kit can be purchased at $195 with additional attachments at $34.95 [9]. A study by The American Journal of Occupational Therapy showed that participants preferred the Liftware Steady™ and weighted spoon when compared to other products. However, between these two utensils there was no statistical difference in ratings from participants [10]. A recent PD device attempts to control tremors through a sensorimotor feedback loop [11]. The feedback loop is used to interrupt the feedback loop causing the tremors. Another study had six patients undergoing adaptive out-of-phase stimulation therapy. This study resulted in a significant decrease in tremor amplitude (67 U + 00B1 13%) in all but one participant [12].

3.2 Intro and Previous Work

The aforementioned assistive devices all sought to decrease the severity of PD tremors. These tremors normally occur within the range of 4–6 Hz. Furthermore, tremors can be a symptom of many disorders, the range of frequencies for most tremors was found to fall between 3 and 12 Hz [8]. Because of this wide range, it is essential that assistive devices effectively suppress this range of frequencies.

This research is an extension of studies on a series of PD assistive devices. The earliest versions involved a utensil designed to suppress tremors while eating. These utensils were initially effective at 5 Hz and then were developed for human subject testing [13, 14]. The work discussed below is directly subsequent to the research done by LeBar, Prentice, and Doughty. In their previous work, an adjustable design was able to suppress tremors from 4 to 7 Hz [15].

These previous devices are unique attempts on suppression PD tremors because of their noninvasive, purely mechanical nature. This style of solution stands out among others in several ways. Most solutions involve active or proactive suppression; however, this style of device can passively suppress tremors. While both methods are capable of suppressing tremors, passive suppression provides additional convenience and novelty by reducing motion in both the hand and the device. Additional features of this solution style are its affordability and effectiveness.

The goal of this research was to incorporate designs from previous projects into an adjustable device. Thus, the same design can effectively suppress a spectrum of tremors, ranging from 4 to 12 Hz. Balancing between the effectiveness and adjustability of the device were essential to the design as those two criteria provided functionality (Table 3.1). Furthermore, additional criteria were determined such that the design was lightweight, low-cost, inconspicuous, and durable.

3.3 Methods

To test each device in a controlled and repeatable environment, a mechanism was constructed to replicate the dynamic behavior of the human hand (Fig. 3.1). System parameters of mass, stiffness, and damping were made to be consistent with those found in a typical human hand using traditional system identification methods. The mass used for the hand was 400 g, and the stiffness and damping were made to be adjustable through the slotted design of the attachment to the rotational base. Once in place, these parameters were verified through a measurement of the mechanical hand's natural frequency of 5 Hz (U + 00B1 0.2 Hz). This is consistent with previous research which describes the natural frequency of a human hand [8].

To run these tests, a shaker was attached to the system and stimulated with a sweeping sine waveform from a LabVIEW code. The code had settings available to change the initial/final frequencies and the number of steps the shaker would take. The standard test applied ran from 2 to 10 Hz with 0.5 Hz step, allowing for a full dynamic characterization of the hand and

Table 3.1 Design criteria

Criteria	Weight	Description
Effectiveness	Essential (5)	The device is effective at suppressing tremors when compared to other solutions. It is effective for frequencies ranging from 4 to 12 Hz
Adjustability	Essential (5)	Every Parkinson's patient is different. The device is meant to help as many people as possible, so the device is designed such that the user can adjust two key features: The effective frequency of the device and the fit of the device on their hand
Health & Safety	Essential (5)	The device is safe for all users and noninvasive
Mass	Important (4)	The device is under 450 g
Aesthetics	Important (4)	The user is confident while wearing the device. It does not draw attention to itself while still being aesthetically pleasing. The device should be small to help attain this goal
Ergonomics	Prominent (3)	The device fits and feels comfortable on the user's hand
Cost	Desired (2)	The cost of the device is competitive, relative to other solutions
Durable and sustainable	Optimal (1)	The device is designed to last a long time and through many climates. Parts should be easily replaceable if something breaks. Materials are environmentally friendly and recyclable

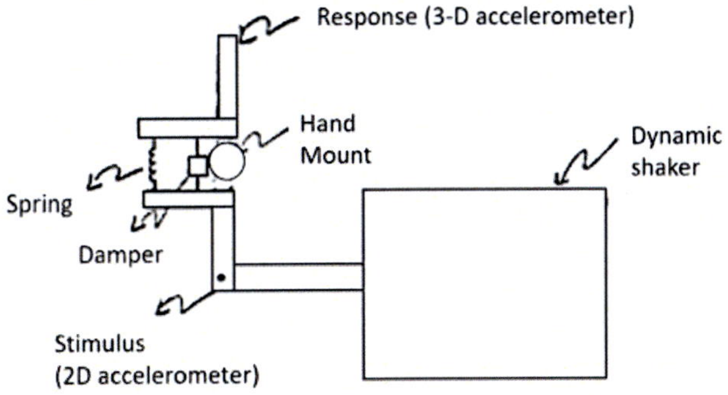

Fig. 3.1 Testing setup top view [14]

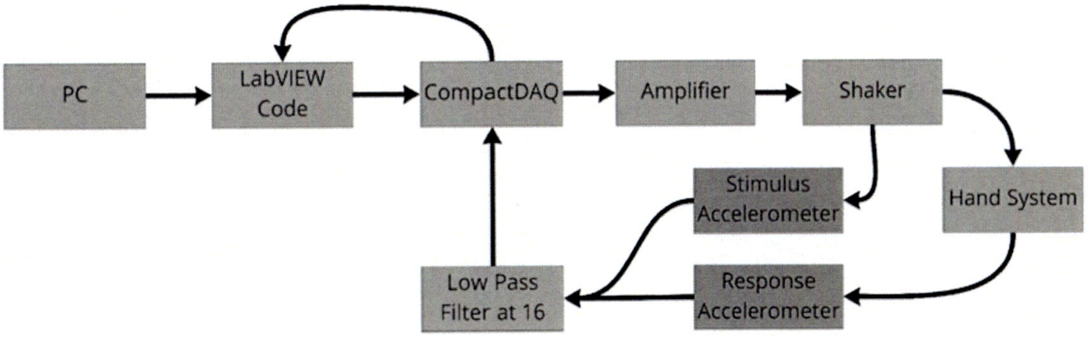

Fig. 3.2 Block diagram of testing setup

device. As the test ran, two accelerometers on the system were used to gather the stimulus waveform generated by the shaker and the response waveform in the hand. Both sets of accelerometer data were sent through a lo-pass filter of 16 Hz and then sent back to the LabVIEW code. From the code, a Frequency Response Function (FRF) could be generated. A block diagram of the test setup can be seen in Fig. 3.2.

When testing the device on the hand, the device was adjusted to target a specific frequency associated with hand tremors. Once the test ran, the Frequency Response Function (FRF) graph generated from LabVIEW showed the device's effectiveness at the desired frequency. After some postprocessing in MATLAB, the following results were produced.

3.4 Results

The following results were generated via MATLAB by overlaying the FRF of the hand while wearing the device and while not. A legend is provided with each FRF to differentiate between the responses. All tests were run from 2 to 15 Hz in 0.2 Hz steps and preformed for various device settings. To test the adjustability of the devices, each was adjusted to three different configurations and analyzed. The effective frequency was determined by the lowest magnitude point on the FRF with the device. To solve for the ratio of reduction, the frequency response (in dB) at the effective frequency of the device was compared to the free hand's frequency response at that same effective frequency. The effective range was determined to be the range of frequencies the device had a ratio of reduction greater than ten (Table 3.2).

3.5 Discussion

The String and The Pin designs were effective at suppressing tremors over the specified range and best met the criteria determined prior. The String design had an overall effective range of 4.7–11.5 Hz whereas The Pin was only effective from 3.2 to 6.2 Hz (Table 3.2). The design criteria outlined the acceptable effective range to be 4–12 Hz (Table 3.1). With this criterion, The String design is more effective than The Pin design. However, The Pin had the highest ratio of reduction in any of the tests. This was most likely because the effective frequency was practically equal to the resonant frequency of the hand, meaning that the difference in magnitudes was maximized (Fig. 3.3a, b). A closer look at the plots will also show that the frequency response of the hand-device system during that test is not any lower than the minimums of other tests.

The String design was also effective over a greater breadth near the frequency to which it was adjusted. This is beneficial for patients when their tremor changes frequency, so the device can still effectively suppress tremors. This will allow them to spend more time doing daily life activities and less time fine-tuning the device before it works. The current designs do not allow for adjustability for comfort; however, both have proven they are adjustable regarding effective frequency.

Table 3.2 Summary of graphical results

Device name	Effective frequency [Hz]	Ratio of reduction at lowest response	Effective range [Hz]
String design	6.06	92.1	4.7–7.8
String design	8.91	25.0	5.0–9.6
String design	11.1	13.5	10.6–11.5
Pin design	3.21	13.3	3.2–3.3
Pin design	5.66	359	5.1–6.2
Pin design	8.91	8.50	N/A

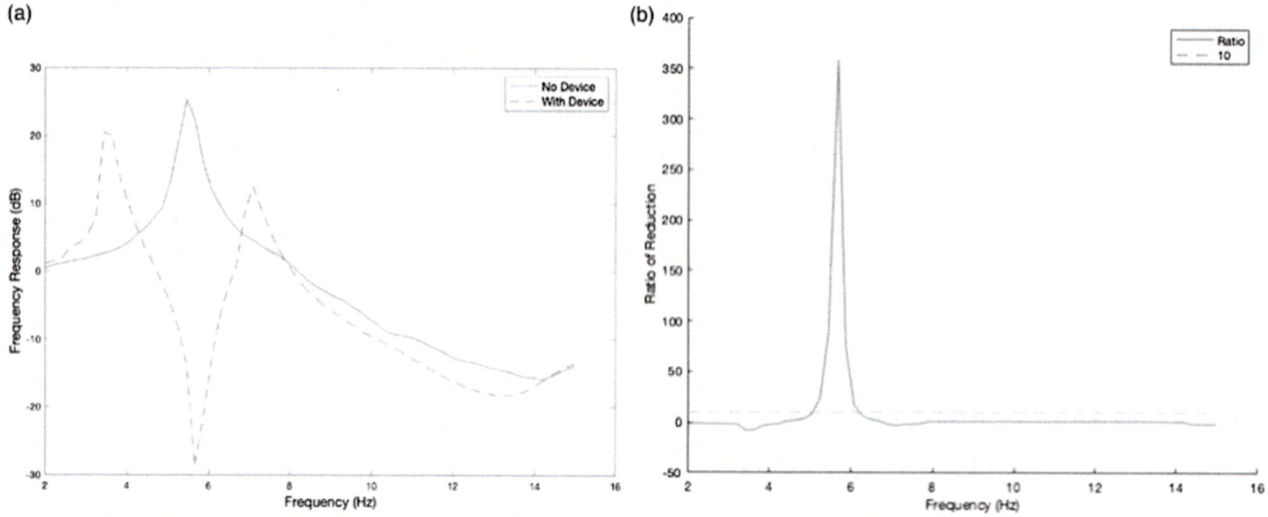

Fig. 3.3 (**a**) Pin design adjusted to middle frequency setting (**b**) Associated ratio of reduction

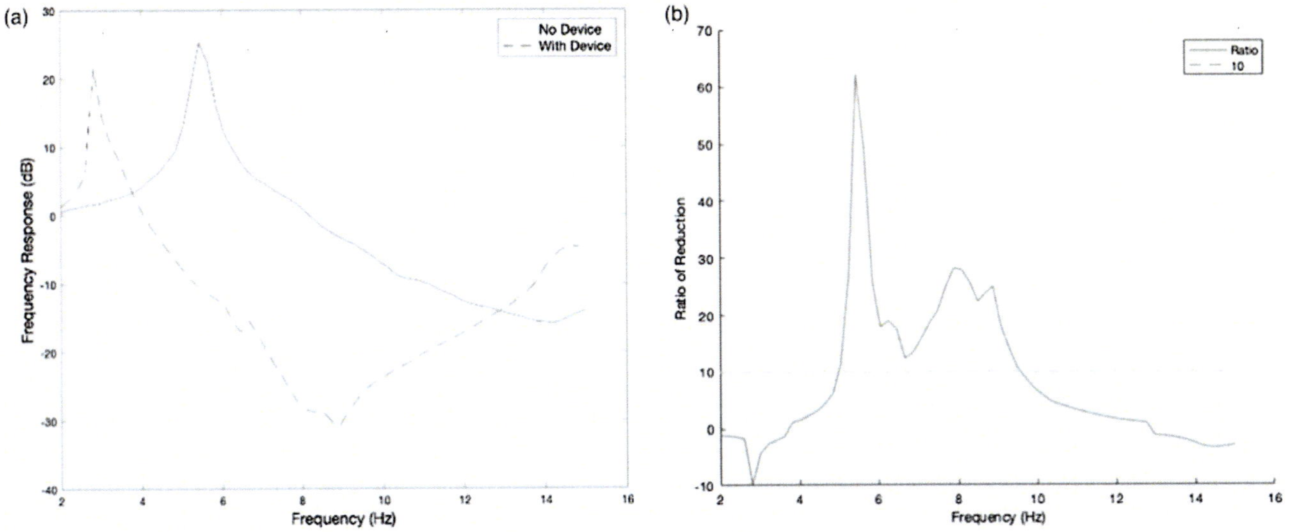

Fig. 3.4 (**a**) String design adjusted to middle frequency setting (**b**) Associated ratio of reduction

When adjusted to its middle setting, The Pin was able to suppress 5.66 Hz effectively with a range of U + 00B1 0.55 Hz (Fig. 3.3a, b).

When adjusted to its middle setting, The String design could effectively suppress 8.91 Hz with an improved span of U + 00B1 2.3 Hz (Fig. 3.4a, b). Both designs struggle with precise and easy adjustability. However, The String design always had a wider effective range, so it would need to be adjusted less often and with less precision. Because of this, The String design preforms better than The Pin with regards to the adjustability design criterion.

Both devices are made from 3D-printed ABS plastic, steel, and silicone so users can wear the device for extended durations. The String is noticeably heavier than The Pin, making for a cumbersome design and therefore failing the mass criteria (515 g and 164 g, respectively). Both devices are built to remain effective for years, even if the individual's tremors change frequency due to the adjustable nature of the designs. With respect to the design criteria, these devices both fail in areas of aesthetics and ergonomics. Both devices are quite conspicuous and cumbersome for a user to have on the back of their hand. The Pin is especially conspicuous due to the 12 cm member coming off the device. Both the devices fit to the hand with a flat 3D-printed plate and no padding. When worn for an extended period, this can lead to discomfort and ultimately cause the user to remove the device. The device must be comfortable to wear so the user can wear it all day.

3.6 Conclusion

Each device comes with various benefits and drawbacks; however, the key feature of both designs is their adjustability. This is a novel development in PD devices which makes these two designs stand out among others. Additionally, the devices have a noticeably lower cost than their competitors, which will significantly increase marketability and accessibility. Some drawbacks of the two designs are in their aesthetics and ergonomics. The Pin design has awkward geometry while the String design is heavier than desired. Both of these issues are currently in study.

These designs serve as proof-of-concept and need to be constantly iterated/tested as they progress through the design process. Moreover, these devices must go through human subject testing as a part of the design process. This will serve to further prove the efficacy of the device and allow Parkinson's patients to give feedback on the design.

In addition to human subject testing, some specific avenues to explore for this research involve material selection, assembly methods, and a possible companion smartphone app. Simple changes in material selection for central parts of each device could vastly improve their shortcomings in both effectiveness and aesthetics. Certain materials were not available during the fabrication phase of this research which forced the use of suboptimal materials. These restrictions also impacted the method in which the devices were assembled. Improving the construction of the designs would also fix the inefficiencies in each device.

A possible long-term goal would be incorporating a companion smartphone app to pair with the device. The various sensors inside smartphones could allow for some useful features. One helpful feature could be using the accelerometers inside smartphones to sense the user's current tremor and give directions on how to adjust their device to improve its effectiveness.

Another feature could involve offering ways to help the mental health challenges that come with PD. For instance, the app could provide helpful resources and information, perhaps by partnering with current charities which already have this content complied and available.

Acknowledgments Special thanks to Jacob Amos, Christina Chrestatos, Allen Hansen, Jared Rees, and Mirza Gurda for all their technical support and guidance. Additional thanks to the Shiley School of Engineering for its essential funding of this research. Finally, thank you to the team's advisor, Dr. Timothy Doughty for his counsel and insight on assistive devices.

References

1. Zach, H., Dirkx, M., Bloem, B.R., Helmich, R.C.: The clinical evaluation of Parkinson's tremor. J. Parkinsons Dis. **5**(3), 471–474 (2015)
2. Marras, C., Beck, J.C., Bower, J.H., Roberts, E., Ritz, B., Ross, G.W., Abbott, R.D., Savica, R., Van Den Eeden, S.K., Willis, A.W., Tanner, C.: Prevalence of Parkinson's disease across North America. NPJ Parkinsons Dis. **4**(21) (2018)
3. Stroke.: March 2020. https://www.ninds.nih.gov/Disorders/Patient-Caregiver-Education/Hope-Through-Research/Parkinsons-Disease-Hope-Through-Research#whatis
4. Mental & Physical Health: Fatigue & Sleep.: The Michael J. Fox Foundation. https://www.michaeljfox.org/news/fatigue-sleep
5. Mental & Physical Health: Diet & Nutrition.: The Michael J. Fox Foundation. https://www.michaeljfox.org/news/diet-nutrition
6. Mental & Physical Health: Depression & Anxiety.: The Michael J. Fox Foundation. https://www.michaeljfox.org/news/depression-anxiety
7. Grissinger, M.R.: Delayed administration and contraindicated drugs place hospitalized Parkinson's disease patients at risk. Pharm Therap. **43**(1), 10–11, 39 (2018)
8. Charles, P.D.M., Esper, G.J.B., Davis, T.L.M., Macinunas, R.J.M., Robertson, D.M.: Classification of tremor and update on treatment. Am. Fam. Physician, 1565–1572 (1999)
9. Liftware.: Verily. https://www.liftware.com/. Accessed Jul 2020
10. Sabari, J., Stefanov, D.G., Chan, J., Goed, L., Starr, J.: Adapted feeding utensils for people with Parkinson's-related or essential tremor. Am. J. Occup. Ther. **73**(3) (2019)
11. Nawrat, A.: Emerging Technology. Medical Technology. https://medical-technology.nridigital.com/medical_technology_apr19/taming_tremors_medical_devices_for_parkinson_s_disease
12. Maneski, L.P., Jorgovanovic, N., Ilic, V., Dosen, S., Keller, T., Popovic, M.B., Popovic, D.B.: Electrical stimulation for the suppression of pathological tremor. Med. Biol. Eng. Comput. **38**(49) (2011)
13. Doughty, T., Bankus, N.: Mechanical modeling and design for reduction of parkinsonian hand tremor. Am. Soc. Mech. Eng. **2**, 521–527 (2010)
14. Doughty, T., Heintz, J., Ishii, M.: Reducing parkinsonian hand tremor with a novel dynamic eating utensil. In: ASME International Mechanical Engineering Congress and Exposition, vol. 3B, (2013)
15. LeBar, K., Prentice, N., Doughty, T.: Design of Purely Mechanical Device to Aid in the Suppression of Parkinsonian Hand Tremors, vol. 5. Society of Experimental Mechanics (2021)

Chapter 4
Fractional Viscoelastic Modeling Enabling Accurate Atomic Force Microscope Contact Resonance Spectroscopy Characterization

Md Tasmirul Jalil, Rafiul Shihab, and Ryan Tung

Abstract In this work, an Atomic Force Microscope (AFM) technique known as Contact Resonance (CR) Spectroscopy is used to measure the complex modulus, as a function of frequency, of a Delrin™ (Polyoxymethylene) sample. These CR experiments, along with AFM creep and relaxation experiments, are conducted at different temperatures, and a time temperature superposition scheme is developed and applied to construct storage and loss modulus master curves over a wide range of frequencies. Large arrays of classical, integer-order, differential equation-based viscoelastic models are typically used to extract the material parameters from the master curves, however, the resulting models exhibit ringing phenomena, poor extrapolation properties, and are numerically cumbersome due the large number of model parameters that are needed. To avoid these pitfalls, we apply fractional viscoelastic models to describe both the time and frequency dependent experimental data, and extract the corresponding mechanical parameters. Fractional viscoelastic models are based on differential equations with fractional derivatives. These models require fewer fitting parameters, compared to their integer-order counterparts, and naturally capture the power-law time domain responses observed in the Delrin™ material.

Key words Atomic force microscopy · Contact resonance · Fractional calculus · Time temperature superposition · Master curve

4.1 Introduction

Bulk properties of viscoelastic materials differ extensively from their nanoscale properties, and quantifying the nanoscale behavior of the material presents numerous challenges both in analytical modeling of the material and experimental measurement techniques. Atomic Force Microscopy is an essential tool used to probe the nanoscale mechanical properties of materials [1]. AFM is capable of quantifying the mechanical response of a wide variety of materials using quasi-static (i.e. force spectroscopy) and dynamic methods (i.e., force modulation microscopy and multifrequency intermittent contact modes) [2].

Contact Resonance Atomic Force Microscopy (CR-AFM) is a sub-method of dynamic AFM where the nanoscale dynamic properties (storage and loss modulus, and loss tangent) of polymers can be extracted with high precision. An AFM cantilever, vibrating freely at any of its natural frequencies will experience a change in both natural frequency and quality factor if it is brought into contact with a viscoelastic material. The shift in the frequency depends on the stiffness of the material, and the change in quality factor depends on the damping characteristics of the material. With a suitable contact mechanics model, the shift in contact natural frequencies and quality factors can be utilized to estimate the nanomechanical viscoelastic properties of the material under inspection [3]. However, the choice of viscoelastic model used in the CR-AFM analysis plays a pivotal role in determining the dynamic properties of the material. Large arrays of classical, integer-order, differential equation-based viscoelastic models (e.g., the generalized Kelvin–Voigt model) can be used to extract the material parameters, yet the resulting models exhibit ringing phenomena, poor extrapolation properties, and are numerically cumbersome due to the large amount of model parameters. On the other hand, modeling viscoelastic materials using the so called "fractional" element offers several advantages to the standard integer-order spring and damper elements previously mentioned [4]. For an integer-order linear spring element, the force produced by the element is proportional to the zeroth derivative of the element displacement. The zeroth derivative represents the identity operator. For an integer-order damper, the force produced by the element is

M. T. Jalil · R. Shihab · R. Tung (✉)
Department of Mechanical Engineering, University of Nevada, Reno, Reno, NV, USA
e-mail: mjalil@nevada.unr.edu; rhshihab@nevada.unr.edu; rtung@unr.edu

A. Amirkhizi et al. (eds.), *Challenges in Mechanics of Time Dependent Materials, Mechanics of Biological Systems and Materials & Micro-and Nanomechanics, Volume 2*, Conference Proceedings of the Society for Experimental Mechanics Series,
https://doi.org/10.1007/978-3-030-86737-9_4

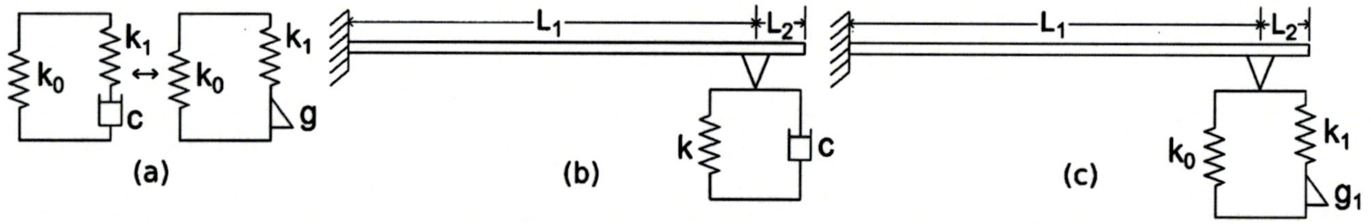

Fig. 4.1 (**a**) The classical damping element c in a SLS model (left), the classical damping element is replaced by a fractional damping element, resulting in a fractional SLS model, (**b**) An Euler–Bernoulli cantilever beam with a classical Kelvin–Voigt element at the tip. The complex characteristic equation of this model is used to determine in-contact frequencies to develop the fictitious cantilevers. (**c**) An Euler–Bernoulli cantilever beam with a fractional SLS element at the tip. The complex fractional characteristic equation derived from this model is utilized to extract fractional SLS model parameters used to predict the material's dynamic response over a wide range of frequencies

proportional to the first time derivative of the displacement. For a fractional element, the force produced by the element is proportional to the pth derivative of the displacement (with respect to time). Where p is a real number between 0 and 1 called the fractional power. For instance, Fig. 4.1a shows a classical Standard Linear Solid (SLS) viscoelastic model consisting of a spring (k_0) in parallel with a Maxwell element (a spring k_1 in series with a damping element c) can be transformed into a fractional SLS model by changing the linear damping element c into a fractional damping element g. In this study, the goal is to describe the dynamic properties of polymers (i.e., storage and loss modulus) over a wide range of frequencies—in other words to form a frequency dependent complex modulus Master curve using CR-AFM and a fractional viscoelastic model. A master curve is a representation that allows the prediction of material behavior on a very large time and temperature scale, a scale that is beyond which is typically measurable [5].

4.2 Concept Validation

To construct a complex modulus master curve using CR-AFM and a fractional viscoelastic model from experimental measurements, the theory is validated by creating a set of fictitious cantilevers that represent possible experimental measurements obtained via CR-AFM at fixed contact resonance frequencies. Delrin™ (polyoxymethylene) is used as the test sample, and the bulk scale storage (E') and loss (E'') modulus data of Delrin™ over four decades of frequency is taken from the literature. Fictitious cantilevers are generated such that the in-contact resonance frequencies for each cantilever-sample case will fall within this frequency range of interest. From the elastic solution for the Young's modulus, using nanoindentation and the elastic-viscoelastic corresponding principle, the tip-sample stiffness ($k = 2E'\sqrt{\Delta/\pi}$) and damping ($c = 2E''/\omega\sqrt{\Delta/\pi}$) [6, 7] are determined from the obtained storage and loss modulus data for a corresponding angular frequency ω, assuming a tip-sample contact area Δ. Contact stiffness and damping parameters are nondimensionalized in terms of the geometric and material parameters of each of the fictitious cantilevers as $\alpha = kL_1^3/3EI$, and $\beta = c\sqrt{L_1^2/9EI\rho A}$. Here, L_1 is the effective length of the cantilever beam (measured from the fixed end to the tip of the beam), E is the Young's modulus of the beam, I is the area moment of inertia ($I = bh^3/12$, where b is the width and h is the thickness of the beam), ρ is the density of the beam, and A is the beam cross-sectional area. Next, the in-contact natural frequencies (f_n^c) and quality factors (Q_n^c) for each fictitious cantilever-sample system are determined by solving the characteristic equation (Eq. 4.2 in Yuya et al. [7]) obtained for an Euler–Bernoulli beam model satisfying the boundary conditions (a classical Kelvin–Voigt element at the tip) and continuity conditions (Fig. 4.1b). Here, the viscoelastic properties of the sample are represented by a complex wavenumber $\lambda_n L_1 = (a_n + ib_n)$, where $a_n = 1.8751\sqrt{f_n^c/f_n^0}$ and $b_n = a_n/4Q_n^c$ (assuming no intrinsic damping of the cantilever) are the real and imaginary part of $\lambda_n L_1$, respectively [2], and f_n^0 are the free natural frequencies of the cantilever. These wavenumbers are calculated by solving the characteristic equation of the cantilever-sample system.

Next, an analytical model of an AFM cantilever with a fractional Standard Linear Solid (SLS) element at its tip (Fig. 4.1c) is derived in terms of fractional parameters and fictitious cantilever properties. The characteristic equation of this model in functional form is:

$$f(L_1, b, h, E, I, \rho, A, \text{contact}\,\lambda_n L_1, k_0, k_1, g_1, p) = 0. \tag{4.1}$$

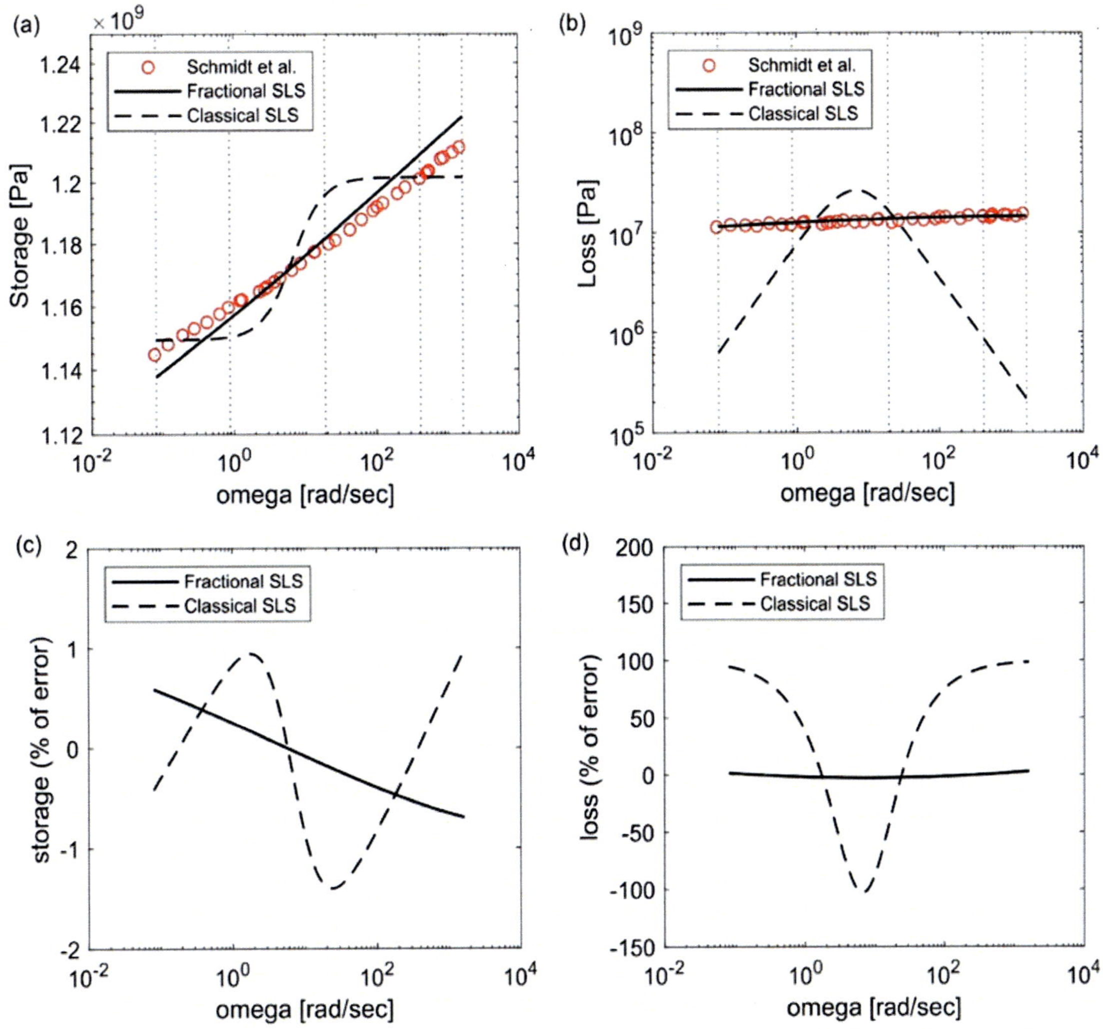

Fig. 4.2 (**a**) Storage and (**b**) loss modulus master curve reconstructed with 5 fictitious cantilevers over four decades of frequency. The red circles represent bulk scale experimental data obtained from Schmidt et al., and the solid and dashed line represent the fitted data generated by a fractional SLS model and a classical SLS model respectively. Percent error in the fractional SLS model (solid) and the classical SLS model (dashed) are shown in (**c**) for storage and (**d**) for loss modulus. Vertical lines in (**a**) and (**b**) show the contact natural frequencies of the fictitious cantilevers used in the analysis

This characteristic equation defines the contact natural frequencies of the system, given the input parameters. For five fictitious cantilevers with their respective contact wavenumbers, a system of five nonlinear complex fractional differential equations is formed. The equations are solved using a nonlinear least-square approach in MATLAB, and the fractional SLS parameters (k_0, k_1, g_1, p) are extracted. Figure 4.2a, b show the storage and loss master curve, generated by using the fractional SLS model, and its classical counterpart with the extracted parameters. The response of the fractional SLS model is in good agreement with the master curves obtained from Schmidt et al. [4].

4.3 Experiment and Future Work

Using a single AFM cantilever would result in a small, finite number of specific shifted CR contact natural frequencies and quality factors (a single shift for each mode) and therefore would be capable of predicting moduli only at those specific frequencies. However, to generate a master curve, we require many data points at different frequencies. In order to overcome this pitfall, the tip-sample contact stiffness needs to be altered by setting different set point forces during CR measurements. Moreover, the CR experiments can be performed at different temperatures as the tip-sample contact stiffness also changes extensively with respect to the sample surface temperature. However, a time (frequency) temperature superposition (TTS) [8–10] scheme must be applied to the CR measurements at different temperatures so that meaningful moduli can be extracted at a reference temperature that we are interested in constructing the master curve. Time (frequency) temperature superposition (TTS) has been widely explored and implemented by the polymer community for a long time [11]. The crux of the concept can be stated as: time and temperature play equivalent roles in the relaxation process of thermorheologically simple polymers. The viscoelastic response of polymers (i.e. creep, relaxation, complex modulus etc.) extracted from short observation times (high frequency) at higher temperatures is equivalent to long observation times (low frequency) at a lower temperature. These relations can be expressed with Eqs. 4.2 and 4.3 [9] showing that, the storage/loss modulus at a specific frequency ω and temperature T is equivalent to the storage/loss modulus at some shifted frequency ($a_T\omega$) at a reference temperature T_0 if the isotherms are plotted in a log-log scale. a_T is the shift factor that is required to shift the modulus extracted at the experimental temperature horizontally, and b_T is the vertical shift factor:

$$E'(\omega, T) = b_T E'(a_T\omega, T_0), \tag{4.2}$$

$$E''(\omega, T) = b_T E''(a_T\omega, T_0), \tag{4.3}$$

$$a_T = {}^{\tau_T}/_{\tau_{T_0}}. \tag{4.4}$$

Moreover, a_T can be expressed in terms of the characteristic relaxation time τ_{T_0} observed at a reference temperature (T_0) and the characteristic relaxation time τ_T at the current experimental temperature (T) (Eq. 4.4). Therefore, conducting AFM based relaxation experiments and thus determining the corresponding relaxation times (adopting the proper relaxation models) at different temperatures are vital tasks in order to find the required shifting parameters.

We are currently working on developing a protocol to perform a number of force relaxation and CR experiments on different polymer samples using an Asylum Research MFP-3D atomic force microscope with different contact force set points at different temperatures. Our goal is to determine the corresponding shift factors from the force relaxation isotherms, and then shift the frequency dependent moduli (storage and loss modulus—obtained from CR experiments at different temperatures) according to the shift factors in the frequency-modulus log-log scale, and develop a modulus representation for a wider range of frequencies at a single temperature (i.e. the reference temperature the master curve is intended to be constructed). Finally, we are interested in studying the capability of fractional viscoelastic models to capture the viscoelastic response of polymers over a wide range of frequencies with a minimal number of fitting parameters.

4.4 Conclusion

In this study, the prospect of using fractional calculus-based viscoelastic models, to predict the response of polymers across a wide range of frequencies is discussed. The response generated by solving the CR-AFM characteristic equation for a number of different fictitious cantilevers using a fractional standard linear solid (SLS) model shows better congruency with the bulk scale experimental data obtained from literature compared to when an integer-order solid model is used. Moreover, the idea of implementing CR-AFM and nanoscale force relaxation experiments with the aid of a time temperature superposition (TTS) scheme to construct the dynamic modulus master curve, is also presented. However, accurate and reliable CR-AFM measurements are faced with numerous challenges from an experimental point of view. Appropriate measurement of the cantilever spring constant, actuation techniques used to achieve tip-sample resonance, and quasi-static measurements in the presence of adhesive forces on the sample are among a few of the challenges to overcome in CR-AFM measurements. Furthermore, the force relaxation experiments also require extensive attention to determine the exact applied local deformation on the sample. Achieving thermal equilibrium for the experimental setup at different temperature is crucial to minimize the thermal drift of both the cantilever and the sample. The experimental scheme is still under development and this group is concerned with the experimental issues arising during the measurement process. Overcoming the challenges should enable both the AFM and polymer community to explore the capability of CR-AFM in determining nanoscale mechanical properties of polymers over significantly large time or frequency scales.

References

1. Binnig, G., Quate, C.F., Gerber, C.: Atomic force microscope. Phys. Rev. Lett. **56**, 930 (1986)
2. Killgore, J.P., DelRio, F.W.: Contact resonance force microscopy for viscoelastic property measurements: from fundamentals to state-of-the-art applications. Macromolecules. **51**(18), 6977–6996 (2018). https://doi.org/10.1021/acs.macromol.8b01178
3. Hurley, D.C., Killgore, J.P.: Dynamic contact AFM methods for nanomechanical properties. In: SPM in Industrial Applications: Nanomechanical Characterization. Wiley (2013)
4. Schmidt, A., Gaul, L.: Experimental investigation and numerical treatment of viscoelastic materials. In: Proceedings of the Intl. Model Anal. Conf. 3.1 (2008)
5. Chailleux, E., Ramond, G., Such, C., LaRoche, C.: A mathematical-based master-curve construction method applied to complex modulus of bituminous materials. In: Road Materials and Pavement Design, EATA, pp. 75–92 (2006)
6. Christensen, R.M.: Theory of Viscoelasticity, 2nd edn. Academic, New York (1982)
7. Yuya, P.A., Hurley, D.C., Turner, J.A.: Contact-resonance atomic force microscopy for viscoelasticity. J. Appl. Phys. **104**, 074916 (2008). https://doi.org/10.1063/1.2996259
8. Williams, M.L., Landel, R.F., Ferry, J.D.: The temperature dependence of relaxation mechanisms in amorphous polymers and other glass-forming liquids. J. Am. Chem. Soc. **77**, 3701 (1955)
9. Ferry, J.D.: Viscoelastic Properties of Polymers. Wiley, New York (1970)
10. Gurp, M.V., Palmen, J.R.: Time temperature superposition for polymer blends. Rheo. Bull. **67**, 5 (1998)
11. Leaderman, H.: Elastic and Creep Properties of Filamentous Materials and Other High Polymers. The Textile Foundation, Washington, DC (1943)

Chapter 5
A Method for Measuring Displacement and Strain Around a Crack of Rubber Sheets Using Digital Image Correlation

Kengo Fujii, Satoru Yoneyama, Ayaka Suzuki, and Hiroshi Yamada

Abstract This study establishes a method for measuring displacement and strain of rubber materials with large and fast deformations using digital image correlation. Using the proposed method, the behavior of the crack tip, which is important for elucidating the growth of cracks is evaluated. A tensile load is applied to the rubber test piece containing the initial crack, and the state is photographed with a digital camera. The results show that oscillating variations displacement and strain rates near the crack from the start of crack growth to fracture are observed.

Key words Rubber · Crack growth · Digital image correlation · Strain · Crack tip behavior

5.1 Introduction

In recent years, rubber materials have been used for various purposes in many members, and further high functionality and high added value are progressing. However, since almost all rubber products currently in use are durable consumer materials, improving the fracture toughness of rubber materials is one of the greatest concerns. Phenomena such as wear and deterioration caused by fracture are the biggest factors that dominate the reliability of rubber products, but it is very difficult to clarify them. The reasons for this are that it is not enough to handle fracture strength and elongation at break, that the fracture phenomenon is a process of formation and growth of fracture nuclei. Therefore, in considering the fracture toughness and durable life of rubber, it is indispensable to consider the fracture mechanics that deals with the crack growth behavior on the premise of the existence of latent defects.

Research on crack growth in rubber materials has been progressing for some time. As a previous study, Beurrot et al. [1] observed the state of the microstructure during fatigue crack propagation in real time using SEM. The details of the crack tip and the state of propagation were clarified, and in addition to the crack branching phenomenon, the correlation between fatigue characteristics and extended crystallinity was considered. However, the underlying mechanism for crack growth, such as the effect of crack tip behavior, has not been clarified.

It is known that complicated phenomena occur at the tip of the crack in a rubber materials. For example, "stress concentration" that causes large strain locally and "velocity jump" in which the crack growth rate rises momentarily. Therefore, the fracture mechanics of rubber so far has been developed by dealing only with the energy balance of the entire system without dealing with the complicated stress field at the crack tip. However, since the crack tip behavior plays an important role in rubber fracture, the importance of crack tip analysis is clear in elucidating the mechanism of crack growth [2, 3].

DIC (Digital Image Correlation) is a measurement technology that measures the displacement distribution in the in-plane direction from images before and after deformation of the measurement object based on changes in the random pattern of the target surface by image processing [4]. In recent years, it is often used as a useful means for measuring the displacement/strain field at the crack tip because it is very simple and can measure the deformation field in the entire field of view and without contact. However, rubber is a material that undergoes large and high-speed deformation, and its characteristics make the crack

K. Fujii (✉) · S. Yoneyama
Department of Mechanical Engineering, Aoyama Gakuin University, Sagamihara, Kanagawa, Japan
e-mail: c5620146@aoyama.jp; yoneyama@me.aoyama.ac.jp

A. Suzuki · H. Yamada
Elastomer Research Unit, Core Materials Research Department, Bridgestone Corporation, Tokyo, Japan
e-mail: ayaka.suzuki@bridgestone.com; hiroshi.yamada1@bridgestone.com

A. Amirkhizi et al. (eds.), *Challenges in Mechanics of Time Dependent Materials, Mechanics of Biological Systems and Materials & Micro-and Nanomechanics, Volume 2*, Conference Proceedings of the Society for Experimental Mechanics Series,
https://doi.org/10.1007/978-3-030-86737-9_5

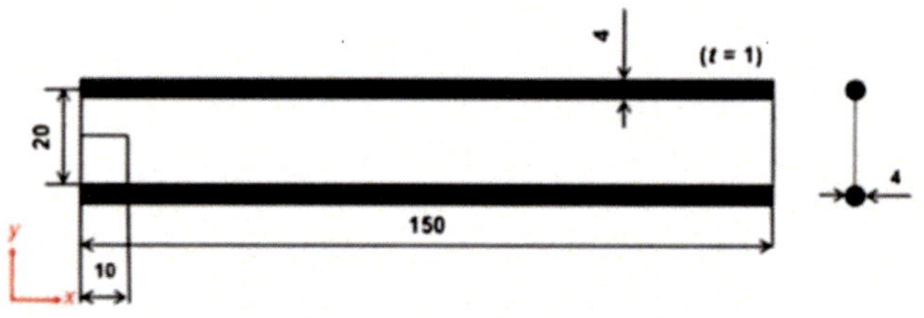

Fig. 5.1 Rubber specimen (mm)

surface extremely large. Therefore, it is difficult to measure and analyze the displacement and strain of the crack tip of the rubber material by the conventional method.

Based on the above, in this study, a method for accurately measuring the displacement/strain field of rubber materials including cracks is established using the digital image correlation method, the crack tip behavior is evaluated using the proposed method.

5.2 Test Method

A tensile load is applied to the sheet-shaped displacement-constrained rubber test piece containing the initial crack. The situation is photographed with a digital camera, and the displacement is measured using DIC from the obtained images before and after deformation. The shape of the test piece used is shown in Fig. 5.1.

The dimensions are length: 20 mm, width: 150 mm, and thickness (= t): 1 mm. A white random pattern is applied to the surface. Also, use a razor to introduce an initial crack of about 10 mm in the center of the left end. A tensile tester is used to load the load, and the tensile direction is downward on the y-axis (negative direction). The test is conducted in room temperature air. The tensile speed and shooting speed are changed for each test, and the analysis range and subset size are changed for each analysis.

5.3 A Method for Displacement and Strain Measurement Around a Crack of Rubber Materials

As mentioned in the previous section, the difficulties of measuring the displacement of the rubber material using DIC are that (1) the deformation is large and fast, and (2) the discontinuity part of the crack expands greatly due to its nature. Regarding (1), under the tensile load of rubber materials, the random pattern suddenly changes due to the large deformation, and the pre-deformation image and the post-deformation image are completely unrelated, so it is difficult to perform measurement. Therefore, images of the deformation process are acquired by high-speed photography, and the reference images are sequentially updated for continuous processing. Then, the displacement is calculated by adding up the obtained results and accumulating them. Regarding (2), Fig. 5.2 shows images before and after deformation of the rubber test piece containing cracks. Figure 5.3 shows the displacement distribution in the x and y directions obtained by a conventional DIC. The displacement distribution is plotted every 10 pixels at the measurement point. From the obtained results, it can be seen that a large error has occurred at the crack. The cause is considered to be the method of treating the discontinuous portion of the crack. As a method for calculating the displacement, an approximate function is created from the rough search result, and the displacement of the entire analysis range is approximately calculated using the function. When the deformation of the cracked portion becomes large, the apparent displacement is displayed in the discontinuous portion. In addition, the calculation including the error of a crack portion requires time for analysis, and further, it is considered that the accuracy of the entire displacement distribution is lowered.

Use the following method to clear the error. In order to calculate without including the discontinuity of a crack, the analysis range is divided into upper and lower parts at the crack part, and the measurement is performed individually. Then, by synthesizing the obtained results, the displacement of the entire analysis range is calculated. The results obtained are shown in Fig. 5.4. It is confirmed that the displacement around a crack can be calculated without causing the error of a crack. In addition, the analysis time can be reduced of the analysis time required to obtain results using the conventional method, and it is considered that the overall accuracy has improved.

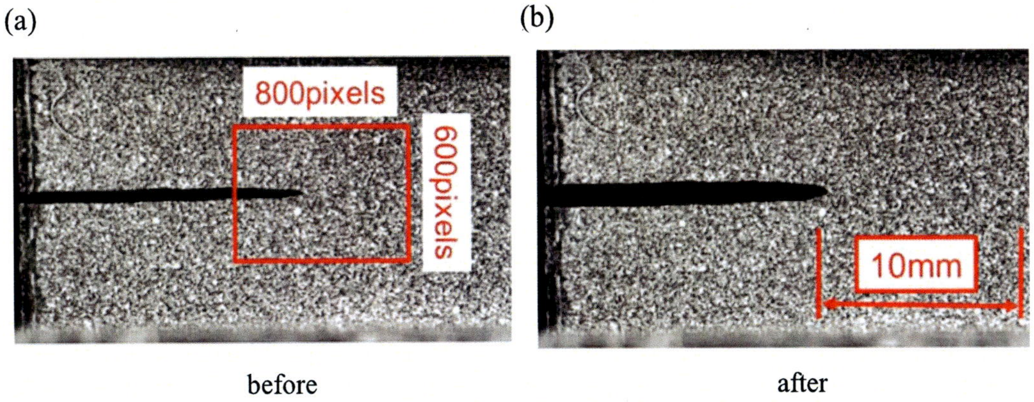

Fig. 5.2 Images before and after deformation

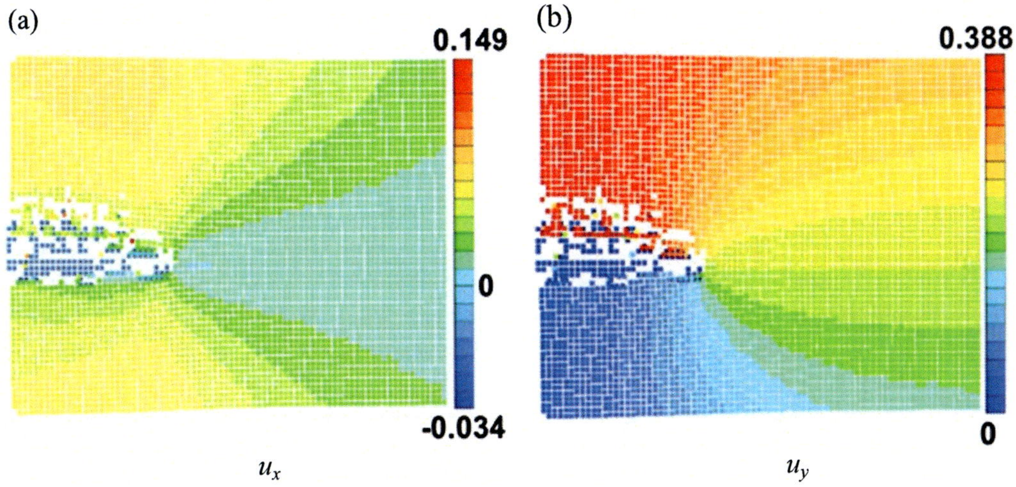

Fig. 5.3 Displacement distribution by the conventional DIC [mm]

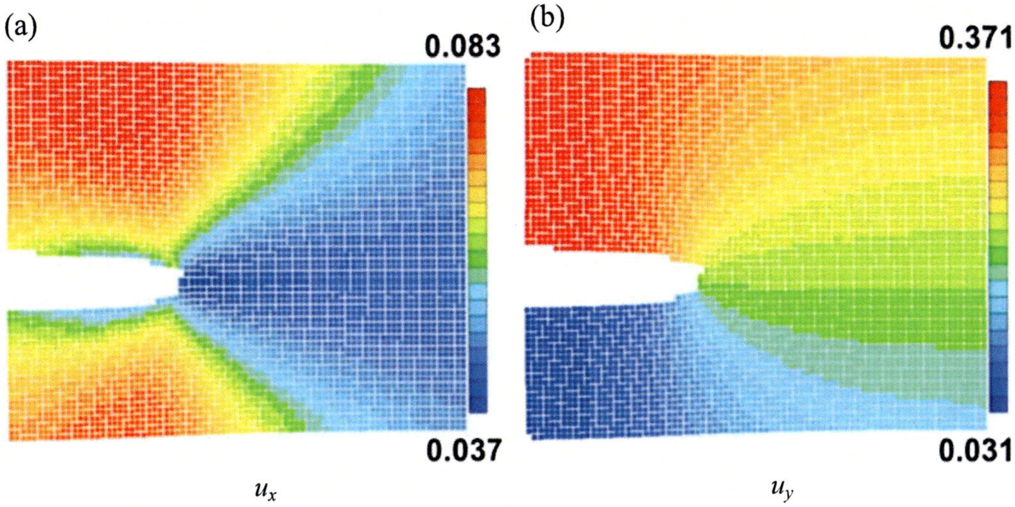

Fig. 5.4 Displacement distribution by the proposed method [mm]

5.4 Evaluation of Crack Tip Behavior During Crack Growth

The method proposed in the previous section is used to measure displacement and strain around crack tip during crack growth. The distance between shots is 1000 mm. The obtained image has a maximum resolution of 1280 × 1024 pixels (8 bits, 1 pixel ≒ 0.022 mm). Tensile speed: 1 mm/s, shooting speed: 200 fps (shutter speed: 3.3 ms), the analysis range is a rectangular range of 250 × 180 pixels (width: 5.5 mm, length: 3.96 mm) including the crack tip, and the subset size is 21 × 21 pixels. Due to the occurrence of velocity jump, it is not possible to capture the state of breakage even at the maximum shooting speed of the high-speed camera, so the image immediately before breakage is taken as the post-deformation image. Figure 5.5 shows the images before and after the deformation, Fig. 5.6 shows the displacement distribution obtained by the proposed method and the logarithmic strain distribution obtained from the displacement calculation results. In addition, the measurement points are plotted every 3 pixels for both displacement and logarithmic strain distribution, and the gauge length is 7 pixels (≒ 0.154 mm). Both of the obtained distributions are smooth, and it can be seen that the shape of the crack tip of the deformed image can be shown.

Based on the above results, it is considered that the proposed method that combines the two solutions that caused the difficulty is appropriate.

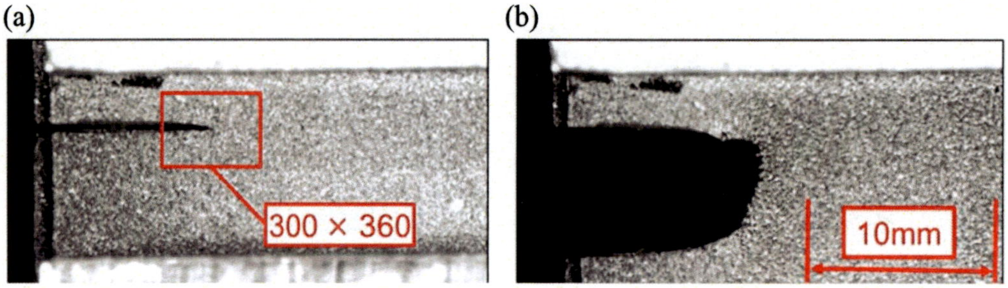

before (b) after

Fig. 5.5 Images before and after deformation

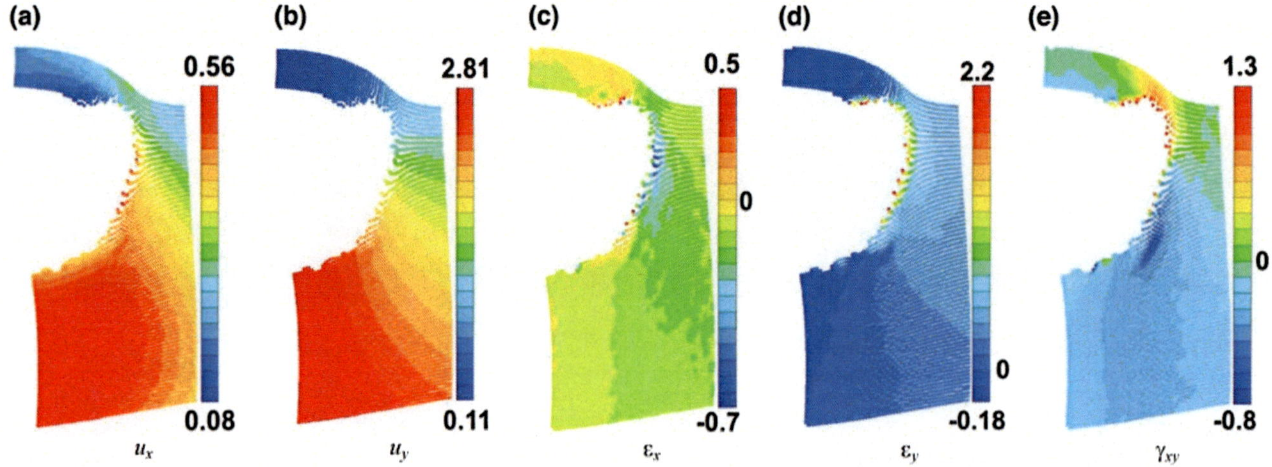

Fig. 5.6 Displacement and logarithmic strain distribution [mm]

5.5 Conclusion

In this study, the crack tip behavior is evaluated in order to elucidate the mechanism of crack growth in rubber materials. As a method for this, a method of measuring the displacement/strain field around a crack tip in consideration of the properties of the rubber material and the existence of cracks using the digital image correlation is proposed. Then, using the proposed method, analysis is performed up to just before fracture, and an accurate displacement/logarithmic strain distribution can be obtained. Furthermore, when the displacement and strain before and after deformation during crack growth are measured at short time intervals, it is confirmed that the amount of deformation had oscillating variations.

References

1. Stephanie, B., Bertrand, H., Erwan, V.: In situ SEM study of fatigue crack growth mechanism in carbon black—filled natural rubber. J. Appl. Polym. Sci. **117**, 1260–1269 (2010)
2. Tsunoda, K.: Fracture mechanics of rubber. J. Soc. Rubber Sci. Technol. Jpn. **87**(5), 168–174 (2014)
3. Fukahori, Y., Sakulkaew, K., Busfield, J.J.C.: Physical aspects for elastic-viscous transition and velocity jump near glass transition region in fracture of rubbers. J. Soc. Rubber Sci. Technol. Jpn. **91**(6), 185–191 (2018)
4. Yoneyama, S.: Basic principle of digital image correlation and measurement procedure of in-plane displacement/strain. Adv. Compos. Mater. **40** (4), 135–145 (2014)

Chapter 6
Understanding the Nanoscale Deformation Mechanisms of Polyurea from In Situ AFM Tensile Experiments

Hanxun Jin, Catherine Machnicki, John Hegarty, Rodney J. Clifton, and Kyung-Suk Kim

Abstract Here, we report in situ AFM tapping-phase study of polyurea's nano and mesoscale phase transitions within the submicron-size field of view. To this end, we designed and assembled a novel in situ AFM loading device that keeps a reference point stationery within the observation window. Using this device, we acquired sequential AFM-tapping-mode phase images of polyurea's nanophase evolution during relaxation under various fixed tensile strains up to 200%. We found that initial hard nano-domains fragment upon rapid loading and the fragmented hard phases go through various nano and mesoscale phase transitions. These fragmented pieces are recombined to form coarsened bicontinuous clusters during the relaxation process. The AFM in situ testing enables us to better understand dynamic-bond characteristics of segmented block copolymers. Interplay between the dynamic-bond characteristics of supramolecular interactions and the hard/soft-phase load/deformation sharing characteristics is believed to predict the self-healing and dynamic toughening mechanisms of polyurea.

Key words Polyurea · In situ AFM tensile test · Nanophase relaxation · Nano and mesoscale phase transitions

6.1 Introduction

Over the past three decades, polyurea has been extensively utilized as a protective coating material owing to its fast reactivity for casting, relative moisture insensitivity, and failure resistance under mechanical loading [1–3]. More recently, polyurea has been further developed as high-toughness nanocomposites to be used under extreme loading conditions [4]. For example, a polyurea copolymer PU1000 was found to have toughness as high as 5500 J/m^2 under a high crack-tip loading rate of $\dot{K} = 10^7\,\mathrm{MPa}\sqrt{\mathrm{m}}/\mathrm{s}$ in a plate impact experiment [4]. The PU1000 has unusually high dynamic toughness compared to other typical polymeric materials that exhibit embrittlement under high strain rate loading; however, so far, corresponding dynamic toughening mechanisms at such a high rate of loading have not been revealed. Molecular and phase-segregated nanostructures of polyurea and their formation processes are relatively well understood [5]. During linear polymerization of polyurea, hard and soft segments self-assemble into unique dual-phase nanostructures consisting of hard arms dispersed in a continuous soft medium, as observed in Atomic Force Microscopy (AFM) tapping-mode phase images [4, 6]. The hard-phase arms have diameters of ~15 nm, and a high-resolution AFM tip of 2 nm radius could reveal details of the nanostructure assemblies through tapping-mode imaging. In contrast, it is not well understood how the partitioning of load and deformation carrying capacities between hard and soft phases and self-healing supramolecular interactions of the nanostructures interplay to generate such a high dynamic toughening effect, spreading inelastic finite deformation without localization at a very short time scale. To get some insights on inelastic deformation mechanisms of the nanophase aggregates, small- and wide-angle X-ray scattering (SAXS/WAXS) caused by nano and microstructural evolution was observed during tensile deformation processes of polyurea [7, 8]. However, the X-ray scattering data primarily provided average hard-domain and intermolecular spacings and their average anisotropic orientations, but could not give information on nanoscale inhomogeneous deformation characteristics of supramolecular interactions and configurational motions of inelastic deformation carriers.

H. Jin (✉) · C. Machnicki · J. Hegarty · R. J. Clifton · K.-S. Kim
School of Engineering, Brown University, Providence, RI, USA
e-mail: hanxun_jin@brown.edu; catherine_machnicki@brown.edu; john_hegarty@alumni.brown.edu; rodney_clifton@brown.edu; kyung-suk_kim@brown.edu

A. Amirkhizi et al. (eds.), *Challenges in Mechanics of Time Dependent Materials, Mechanics of Biological Systems and Materials & Micro-and Nanomechanics, Volume 2*, Conference Proceedings of the Society for Experimental Mechanics Series,
https://doi.org/10.1007/978-3-030-86737-9_6

In effect, we have to trace the nano-structural evolution with a nanometer-scale spatial resolution to uncover the interplay between the hard/soft-phase load/deformation sharing characteristics and dynamic-bond characteristics of supramolecular interactions in a submicron-size observation window. However, we have two significant experimental difficulties. One is drifting of the in situ observation sight due to micron-scale displacements made by specimen deformation and loading-frame motion. The other is the loss of inter-frame image correlations caused by breaking and reforming dynamic-bond characteristics of the supramolecular interactions. Various in situ AFM loading devices have been previously designed to avoid the examination-sight drifting out of the observation window during the loading process, employing either mechanical [9–11] or electrical [12] control schemes. In the mechanical control scheme, typically, a twin-screw composed of left- and right-handed screws of the same pitch were threaded on a single driving shaft and driven by a stepper motor to stretch the specimen symmetrically, keeping the specimen centered in the field of view. The mechanical passive control device allowed observation of specimen deformation or cracking processes approximately near the stationary-point vicinity, within an observation window of 3 μm × 3 μm [10] or 20 μm × 20 μm [9]. However, the devices could not keep inter-frame correlations to trace the same material points. The uncertainty of the displacement that makes the material points move out of the observation window was too large to track the Lagrangian material deformation characteristics. Here, we made a twin-screw-driven in situ AFM loading device and analyzed the uncertainty of the device's material-point traceability by investigating AFM surface-topography's Digital Image Correlation (DIC) within 10 μm × 10 μm windows and AFM tapping-phase's spectral evolution within 1 μm × 1 μm windows. We studied the loading device characteristics by analyzing the DIC of AFM topography images of the sample under several stepping increments of small strains within 10 μm × 10 μm windows. Then, we investigated deformation characteristics of the phase-segregated nanostructures with high-resolution in situ AFM-tapping-mode phase images and their spectral characteristics within 1 μm × 1 μm windows, during stress relaxation at various strain levels up to 200% using the home-made in situ AFM loading device. The nanophase relaxation test removes the uncertainty of the observation-sight motion caused by the loading device operation since the device operation was frozen during the entire relaxation process. The study of transient nanophase relaxation at different strain levels could help understand copolymers' dynamic and self-healing properties.

6.2 Experiments

6.2.1 Formation of Polyurea Thin Films

As shown in Fig. 6.1a, the polyurea was prepared from 4,4′-methylene diphenyl diisocyanate (MDI) (Isonate 143L; Dow Chemical) [13] and polytetramethylene oxide-di-*p*-aminobenzoate (Versalink P-1000; Air Products) [14]. The elastomer is formed through a linear polymerization reaction resulting in a copolymer with urea linkages. These urea linkages, when aligned, form bidentate hydrogen bonds (Fig. 6.1b) forming polar, high-Tg, 'hard' segments; additionally, low-Tg 'soft' segments form from the amorphous poly(tetramethylene oxide) (PTMO). Due to the polar nature of the hard segments and the nonpolar soft segments, the hard segments aggregate to form a micro-phase separation within the long polymer soft segments [7]. The hard regions can be detected with AFM Fig. 6.1c as a bright region while the soft segments are dark [4].

Briefly, Versalink P-1000 was heated to approximately 80 °C or until the viscosity of the oligomer was sufficiently decreased to allow for pouring. This was then added to a 500 mL beaker followed by the rapid addition of MDI in a 1:4 ratio (MDI to P-1000) by weight. Note that MDI was weighed with a 5% excess, as recommended by the manufacturer, to account for a possible side reaction between the diisocyanate moieties and the moisture in the ambient air. The mixture was stirred with a glass rod for 1 min and then degassed until visible bubbles were gone, for a maximum time of 10 min. Following degassing, approximately 5 mL of the curing mixture was added to a mold consisting of a Teflon™ plate (McMaster-Carr, 8711K96) with 200 μm thick tape around the edges for height guidance, producing film dimensions of 2.5 in × 4.5 in × 200 μm. With a glass rod or straight edge, the polyurea was spread across the Teflon™ mold. The casted polyurea cured for 24 h in room temperature conditions. The film was then cut into pieces for the AFM device or in a dog bone shape for tensile testing.

6.2.2 In Situ AFM Loading Device

A home-made in situ AFM loading device driven by twin-screws to self-align observation sights has been designed and assembled. The schematics of the device mounted on a base (1) are illustrated in the left diagram of Fig. 6.2. A left- and right-hand threaded leadscrew (3) with pitch size 0.8 mm is fabricated from a brass rod with a diameter of 4 mm. A stepper motor

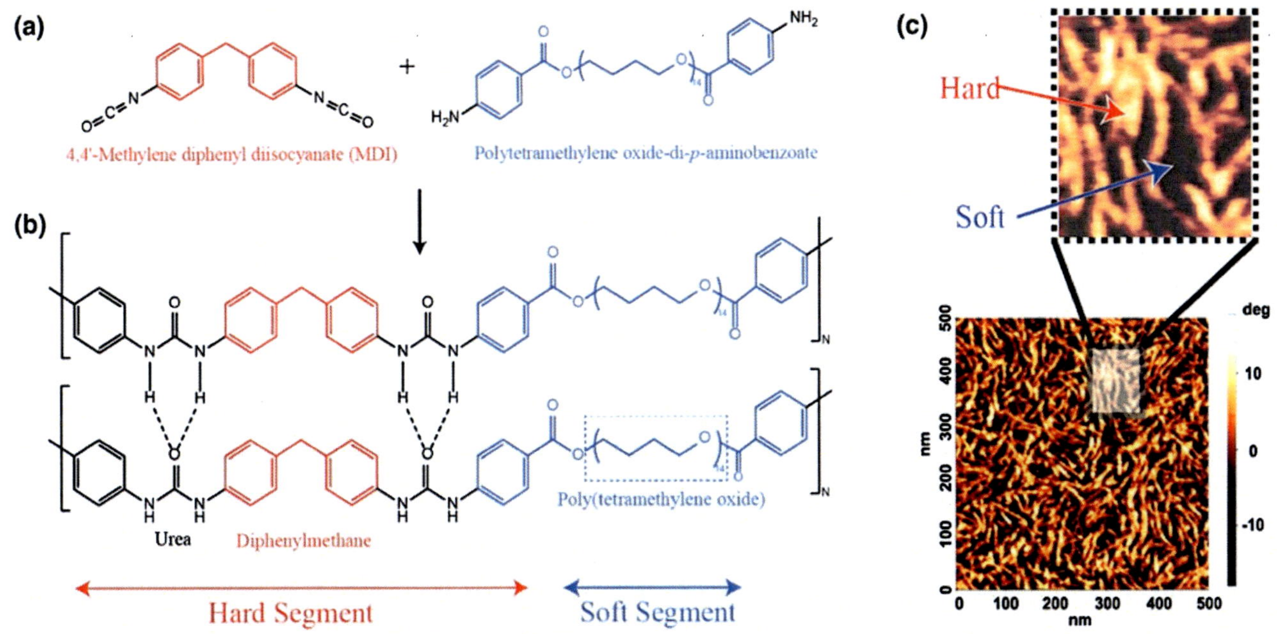

Fig. 6.1 Schematic of polyurea polymerization (**a**, **b**) of 4,4′-methylene diphenyl diisocyanate (MDI) and polytetramethylene oxide-di-*p*-aminobenzoate (Versalink P-1000; Air Products). Chemical structure of bidentate hydrogen bonding between urea linkages to form hard segments. Soft segments arise from the repeating unit of poly(tetramethylene oxide). (**c**) 500 × 500 nm tapping-mode AFM phase images of polyurea indicating bright regions for hard segments and dark for soft matrix [4]

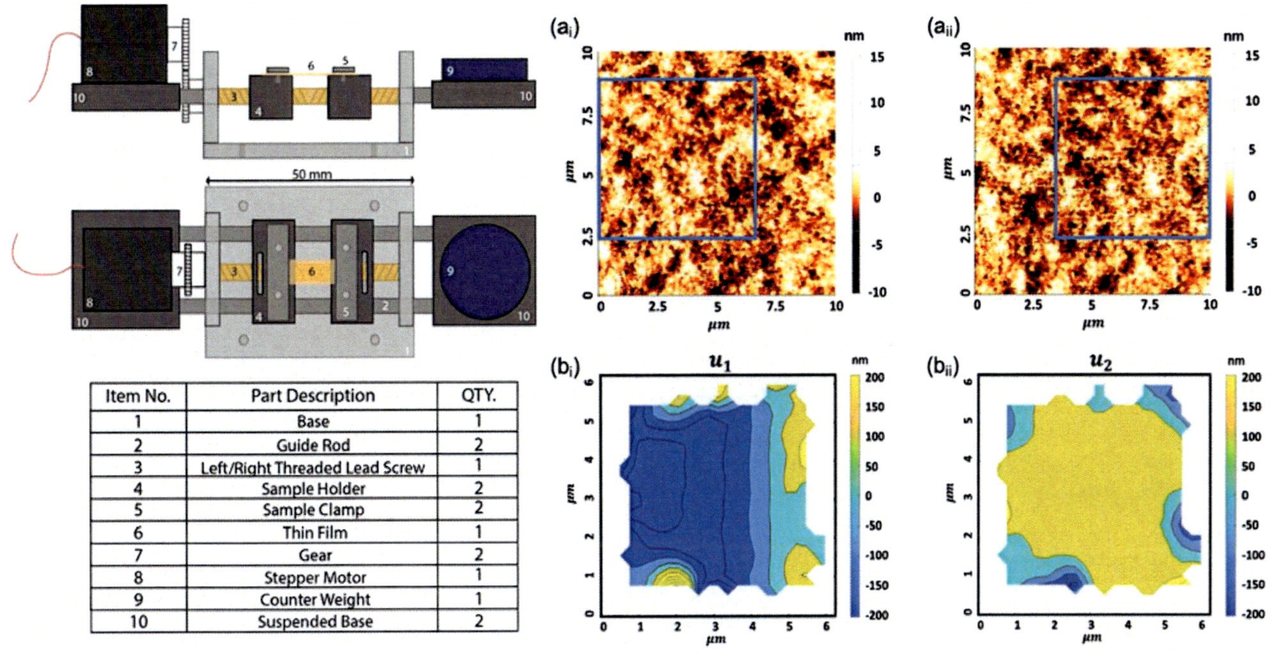

Item No.	Part Description	QTY.
1	Base	1
2	Guide Rod	2
3	Left/Right Threaded Lead Screw	1
4	Sample Holder	2
5	Sample Clamp	2
6	Thin Film	1
7	Gear	2
8	Stepper Motor	1
9	Counter Weight	1
10	Suspended Base	2

Fig. 6.2 Schematics of in situ AFM loading device with observation-sight invariance control and sequential AFM topography images of scan size 10 μm × 10 μm of a PU-1000 sample under tension in the vertical direction with tensile strains, (**a$_i$**) $\varepsilon = 0.32\%$ and (**a$_{ii}$**) $\varepsilon = 0.65\%$. The displacement plots obtained from DIC for axial (**b$_i$**) and transverse (**b$_{ii}$**) displacements

(McMaster-Carr, NEMA) (8) with 200 steps per rotation and operation torque 1.6 in.-oz. is used to precisely control the loading strains by transmitting the torque through one pair of gears (7) with the same teeth number. The motor is controlled by an A4988 motor driver, which is programmed by an Arduino chipboard. The finest strain resolution is calculated as 0.027%/step if the initial sample length is 30 mm. Special beam spring anti-backlash base block (4) has been manufactured to reduce the backlash problem between the dual leadscrew and the sample holders sliding on two guide rods (2). A guide rod is not drawn in the side view to show the twin-screw clearly. A counterweight (9) sits on a holding block (10) to balance the device on the AFM stage. A thin-film sample (6) is fixed to the holders firmly by clamps (5). A cushion block with knurl could be applied to remove potential slip between the sample and the grips under large deformation. The maximum dimensions for the specimen are 30 mm × 10 mm × 1 mm. Our device is flexible to change the dual leadscrew with different pitch sizes and change the pair of gears for different accuracy requirements. In this paper, as the device was primarily used to investigate the device-kinematics control and the sample's displacement fields in the observation window, an optical-gauge load cell was not applied to the device.

6.2.3 *Digital Image Correlation (DIC) Analysis of the In Situ Loading System*

We utilized a DIC of sequential AFM topography images of scan size 10 μm × 10 μm to analyze the observation-sight displacement drifting made by successive loading steps of the stepper motor. We employed a q-factor-based digital image correlation algorithm (qDIC) [15] for the displacement analysis, and the AFM DIC had ~5 nm displacement resolution. Figure 6.2a_i, a_{ii} show two selected frames of AFM surface-topography images before and after the specimen was stretched in two steps of the stepper motor. The end-to-end stretching displacement of the 5 mm long specimen with 5 mm width and 200 μm thickness was 16 μm between the two frames. The maximum square correlation window size was first identified approximately, and one image was translated close to the other image before they were correlated to get the additional fine adjustment of the total correlation displacement. The maximum correlation window size was 6.25 μm × 6.25 μm, and the two well recognizable images are marked with blue boxes. The observation-sight displacement was 0.10 μm in the stretching direction and 3.53 μm in the lateral direction between the two surface-topography images.

6.3 AFM Tapping-Mode Imaging of PU-1000 Nanophases Under Stress Relaxation

The AFM tapping-mode phase and topography images were captured using XE-Bio AFM from Park Systems. The SuperSharpSilicon™ Non-Contact/Tapping-mode probes with reflective coatings (SSS-NCHR) from Nanosensors™ were used. This probe has a curvature radius of 2 nm and a spring constant of 42 N/m. Compared with regular tapping-mode probes with a radius of curvature of 10 nm, SSS-NCHR tip could resolve the nanophases with higher resolution. The set-point ratio was tuned ~0.5 to get high-resolution phase images. The slow-scanning direction was aligned in the tensile direction to avoid vibration noise from the suspended thin film. The AFM tapping-phase image scan size was either 500 nm × 500 nm or 1 μm × 1 μm. Images were taken at a fast-scanning frequency of 2 Hz under ambient conditions, and the image pixel resolution was 256 × 256. Therefore, each image acquisition took roughly 2 min. The controlling motor system was turned off during AFM scanning to avoid any electrical noises.

6.4 Results and Discussions

Regarding the DIC analysis of the in situ loading system, the 0.10 μm displacement of the observation sight in the stretching direction could be caused by the sample's stretching deformation and a 31.3 μm center-position misalignment of the AFM scanning window's center in the stretching direction. However, 3.53 μm displacement in the lateral direction is too large to be made by a Poisson contraction and a center-position misalignment. Such a relatively large lateral displacement is likely made by relative shear and rotation between the specimen holders. The DIC analysis reveals that the twin-screw driving can keep the AFM observation site within the submicron-size window with ~10 nm displacement/step-loading in the stretching direction if the center positioning accuracy of the AFM scanning window is within ~10 μm and the specimen length is reduced to 3 mm. Analysis indicates that a similar control accuracy can be achieved with a slider properly designed to limit the sample holder's rotation. The DIC analysis also showed a sudden jump of observation-sight displacement at a large stretching step, presumably caused by a partial slip of the sample at a frictional grip. Besides, the DIC of AFM surface-topography images kept strong correlations, while the tapping-phase images quickly lost their correlations at a large strain.

Figure 6.3a shows three in situ AFM tapping-mode phase images of fully relaxed nanophases of a thin polyurea film in equilibrium under stretching strains of 0%, 110%, and 200%. We derived the relationship among the tapping-mode phase φ,

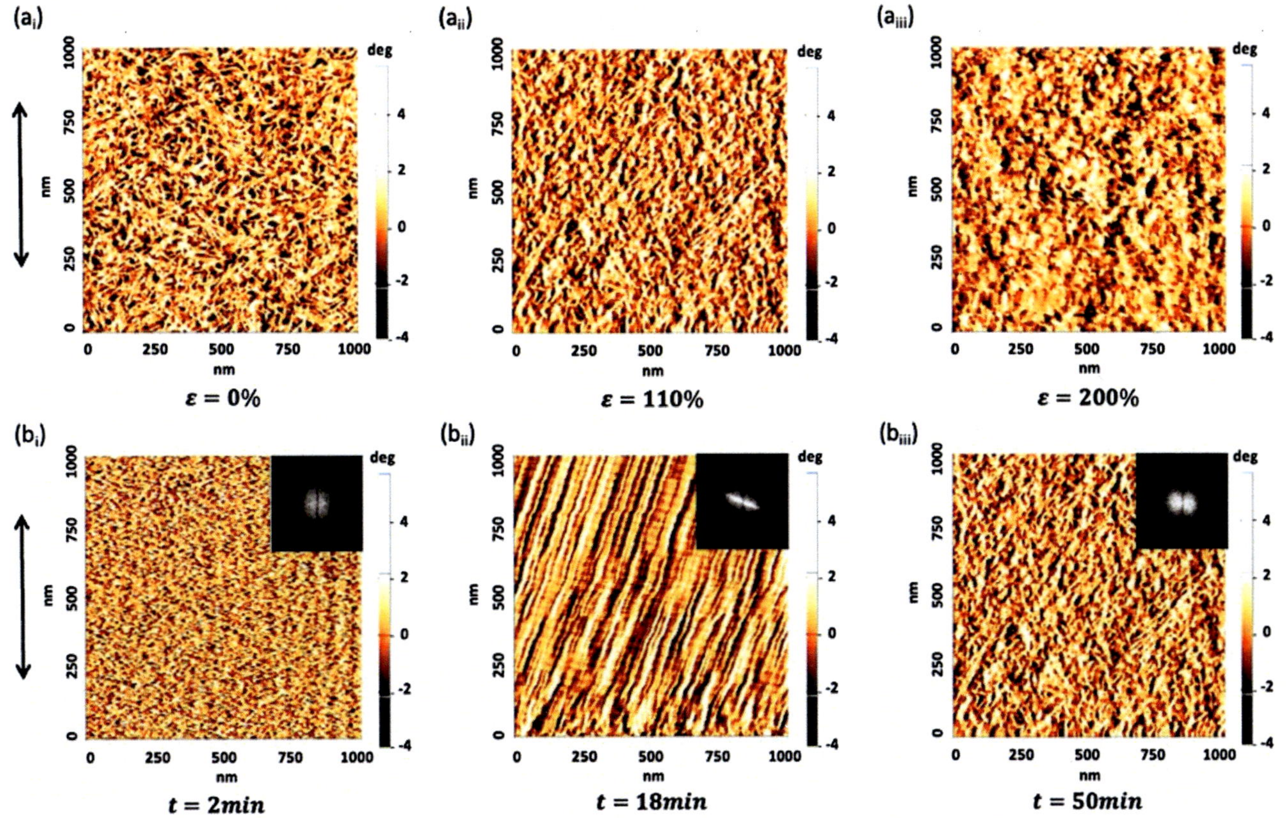

Fig. 6.3 In situ AFM tapping-mode phase images of a polyurea thin film under various strains for **(a_i)** $\varepsilon = 0\%$, **(a_{ii})** $\varepsilon = 110\%$, and **(a_{iii})** $\varepsilon = 200\%$. The time variant AFM tapping-mode phase images during relaxation at $\varepsilon = 110\%$ for **(b_i)** $t = 2$ min, **(b_{ii})** $t = 18$ min, and **(b_{iii})** $t = 50$ min. The FFT plots are inserted in **(b)**. The loading direction is indicated by the black arrow

the work of adhesion w_0, and the elastic modulus E as $\sin\varphi \cong c_1 + c_2 w_0^{7/3} E^{2/3}$. The coefficients c_1 and c_2 only depend on AFM characteristics such as the AFM tip radius, cantilever stiffness, and tapping frequencies, but not on the sample properties. Since the polar, high-Tg, hard segments have larger adhesion and elastic modulus values, the bright regions in the image represent the segregated MDI hard-phase nanostructures. The sample was stretched rapidly, but incrementally in two to three steps, to each strain level, and the strain was held fixed while the stress was partially relaxed, and the nanophases reached equilibrium at the strain level. Each AFM image of Fig. 6.3a was scanned at least 30 min after the sample was stretched to the strain level to ensure the nanophases were fully relaxed to reach the equilibrium at the strain level. As shown in the figure, the nanostructures evolved during the relaxation process, more significantly coarsening the hard phase at larger strains by breaking and recombining the hard domains; in situ WAXS/SAXS study [7, 8] also suggested such nano-structural evolution. Since the stretching strains are very large, the three images' observation sights may not overlap each other but locate within close vicinity, as we used the twin-screw loading device. Figure 6.3a_i shows the AFM phase image of the undeformed nanostructures. The bright regions represent hard MDI domains, which are rod shaped and ~15 nm in diameter and a couple hundred nm in length. The dark region represents the soft diamine group matrix. As the polyurea deformed beyond yielding, the nanophases clustered in equilibrium at each strain level, in addition to the coarsening of individual MDI nanostructures. The clustered patterns are more aligned to the stretching direction at higher strain levels. The alignment is more evident in FFT distributions, as shown in Fig. 6.3b_{iii}. Due to the chemical crosslink agents in Isonate 143L, polyurea tends to stiffen itself at large strain, and strain-induced crystallization may play a role in clustering the hard domains.

Figure 6.3b displays AFM tapping-phase images in the process of PU-1000 nanophase relaxation in 50 min at a fixed stretching strain of 110%. The strain increased from 78 to 110% within 3s, i.e., the strain rate is ~$0.11 s^{-1}$, and the strain was held fixed while the nanophases were relaxed to be in equilibrium at the strain level. The sample was then successively scanned at the same area every 2 min to get AFM tapping-phase images of the nanophases during the relaxation process. Figure 6.3b_i shows the scan image at $t = 2$ min. Interestingly, the inset FFT frequency distribution indicates that the original hard-domain arms were fragmented into smaller rod-shaped segments with the long axis densely aligned perpendicular to the loading direction. As the relaxation process continued, another phase transition occurred at $t = 18$ min. As shown in Fig. 6.3b_{ii}, the short segments were recombined into longer and thicker rods, forming a fibril-bundle structure. These artificial fibril-bundle structures seem to be formed primarily due to the material drifting of polyurea during the relaxation process. After a long relaxation time ($t = 50$ min), the hard MDI domains transformed to coarsened structures, with more domains and their clusters aligned along the loading direction as shown in the FFT image of Fig. 6.3b_{iii} inset. Here, we have shown that many interesting nano and mesoscale phase transitions associated with dynamic-bond characteristics could be discovered with our AFM in situ loading device. We could also quantify the mesoscale phase transitions of a copolymer PU-1000 with tensorial spectral moments of the FFT images. The current multiscale phase transition study is more applicable to understanding self-healing characteristics of complex copolymers [16, 17]; however, the hydrogen-bond fragmentation behavior of the hard domain can give insights to modeling dynamic toughening mechanisms. Detailed experimental and theoretical understanding of these interesting phase transitions combined with mesoscale molecular dynamics simulations will be presented in follow-up publications.

6.5 Conclusion

We designed and analyzed a particular in situ AFM loading device with a twin-thread screw that provides a stationary reference point in the observation window. This device enabled us to collect high-resolution AFM tapping-mode phase images of polyurea nanophases during stress relaxation at various fixed strains ranging from 0 to 200%. To our knowledge, this is the very first in situ AFM tapping-phase study of segmental block copolymers' nano and mesoscale phase transitions within the submicron scan size. We found that initial hard nanostructures fragment upon rapid loading and the fragments undergo various nano and mesoscale phase transitions to reform bicontinuous structures during the relaxation process. The structures relax back to homogenous and isotropic structures below yielding, and the principal axes of the reformed hard domains tend to align along the tensile direction beyond yielding. Strain-induced crystallization may occur and tend to cluster nanostructures at larger strains when the polyurea starts to stiffen. Insights were gained to understand self-healing and dynamic toughening mechanisms of copolymers by studying of fragmentation and reformation processes of hard nanophases in PU-1000, with the home-made in situ AFM loading device.

Acknowledgments This work was supported by the U.S. Office of Naval Research through grant N00014-18-1-2513.

References

1. Barsoum, R.G.S.: Elastomeric Polymers with High Rate Sensitivity: Applications in Blast, Shockwave, and Penetration Mechanics. William Andrew (2015)
2. Grujicic, M., et al.: Experimental and computational study of the shearing resistance of polyurea at high pressures and high strain rates. J. Mater. Eng. Perform. **24**, 778–798 (2015)
3. Kim, K.-S.: Pressure- and rate-dependent plastic flow of nano-phase segregated polyurea copolymer. In: (IMECE Paper # 13559), Drucker Medal Symposium, ASME 2019 IMECE (2019)
4. Kim, K.-S., Jin, H., Jiao, T., Clifton, R.J.: Dynamic fracture-toughness testing of a hierarchically nano-structured solid. In: Fracture, Fatigue, Failure and Damage Evolution, vol. 3. Springer (2021)
5. Holzworth, K., Jia, Z., Amirkhizi, A.V., Qiao, J., Nemat-Nasser, S.: Effect of isocyanate content on thermal and mechanical properties of polyurea. Polymer. **54**(12), 3079–3085 (2013)
6. Magonov, S.N., Cleveland, J., Elings, V., Denley, D., Whangbo, M.H.: Tapping-mode atomic force microscopy study of the near-surface composition of a styrene-butadiene-styrene triblock copolymer film. Surf. Sci. **389**(1–3), 201–211 (1997)
7. Pathak, J.A., et al.: Structure evolution in a polyurea segmented block copolymer because of mechanical deformation. Macromolecules. **41**(20), 7543–7548 (2008)
8. Rinaldi, R.G., Boyce, M.C., Weigand, S.J., Londono, D.J., Guise, M.W.: Microstructure evolution during tensile loading histories of a polyurea. J. Polym. Sci. Part B: Polym. Phys. **49**(23), 1660–1671 (2011)
9. Tambe, N.S., Bhushan, B.: In situ study of nano-cracking in multilayered magnetic tapes under monotonic and fatigue loading using an AFM. Ultramicroscopy. **100**(3–4), 359–373 (2004)
10. Bamberg, E., et al.: A tensile test device for in situ atomic force microscope mechanical testing. Precis. Eng. **30**(1), 71–84 (2006)
11. Deben UK.: In-situ Tensile & Compression stages for SEM. https://deben.co.uk/tensile-testing/sem/mini-tensile-tester-200n-compression-horizontal-bending-stage/
12. Asylum NanoRack™ sample stretching stage. https://afm.oxinst.com/products/mfp-3d-other-driving-forces-acccessories/mfp-3d-nanorack-sample-stretching-stage
13. The Dow Chemical Company.: Isonate 143L. https://www.dow.com/content/dam/dcc/documents/en-us/productdatasheet/109/109-00713-01-isonate-143l-modified-mdi-tds.pdf?iframe=true

14. VERSALINK P-1000 Oligomeric Diamine. https://www.pu-additives.com/en/brands/versalink?xd_co_f=MWUyNzRmMmUtOWU0ZC00NDg0LThjOGItZDJiNTdkNzg5NTI4
15. Landauer, A.K., Patel, M., Henann, D.L., Franck, C.: A q-factor-based digital image correlation algorithm (qDIC) for resolving finite deformations with degenerate speckle patterns. Exp. Mech. **58**(5), 815–830 (2018)
16. Wojtecki, R., Meador, M., Rowan, S.: Using the dynamic bond to access macroscopically responsive structurally dynamic polymers. Nat. Mater. **10**, 14–27 (2011)
17. Ying, H., Zhang, Y., Cheng, J.: Dynamic urea bond for the design of reversible and self-healing polymers. Nat. Commun. **5**, 3218 (2014)

Chapter 7
Porosity Determination and Classification of Laser Powder Bed Fusion AlSi10Mg Dogbones Using Machine Learning

Caroline E. Massey, David G. Moore, and Christopher J. Saldana

Abstract Metal additive manufacturing allows for the fabrication of parts at the point of use as well as the manufacture of parts with complex geometries that would be difficult to manufacture via conventional methods (milling, casting, etc.). Additively manufactured parts are likely to contain internal defects due to the melt pool, powder material, and laser velocity conditions when printing. Two different types of defects were present in the CT scans of printed AlSi10Mg dogbones: spherical porosity and irregular porosity. Identification of these pores via a machine learning approach (i.e., support vector machines, convolutional neural networks, k-nearest neighbors' classifiers) could be helpful with part qualification and inspections. The machine learning approach will aim to label the regions of porosity and label the type of porosity present. The results showed that a combination approach of Canny edge detection and a classification-based machine learning model (k-nearest neighbors or support vector machine) outperformed the convolutional neural network in segmenting and labeling different types of porosity.

Key words Machine learning · Additive manufacturing · Convolutional neural network · Computed tomography · Porosity analysis

7.1 Introduction/Background

Porosity analysis is an important activity in the certification of additively manufactured parts. Current commercial software is available (e.g., Volume Graphics, AVISO) that aids in identifying internal porosity in computed tomography (CT) data. Such software is expensive, and results can vary greatly depending on input conditions, including the quality of the image data and the experience of the software user. There are three types of porosity commonly found in additively manufactured parts: lack of fusion, keyhole porosity, and entrapped gas porosity. Lack of fusion porosity occurs when the power of the laser is insufficient for the laser speed, this leaving large, highly irregular voids in the interior of the part [1]. Gas porosity occurs when shielding gas, melt pool vapor, or powder particles getting stuck in the print layer resulting in a spherical shape [2]. Keyhole porosity is a result of the laser having to slow down as it changes trajectory in the scan track of the part [1]. This can be due to velocity changes at the instances where the laser is changing path, this causing excess energy and evaporation, and leading to a depression in the surface [1]. Keyhole pores are mostly round but not completely spherical [2].

Cunningham et al. [3] states that it is difficult to determine keyhole porosity based on morphological features alone, but by observing the location of the pore and samples with similar processing conditions, one can make this determination. For the purposes of the present study, the authors consider keyhole porosity and gas porosity as spherical porosity. Lack of fusion porosity is believed to be more detrimental to the health of the part than spherical porosity due to its irregular morphology

C. E. Massey (✉)
Sandia National Laboratories, Albuquerque, NM, USA

George W. Woodruff School of Mechanical Engineering, Georgia Institute of Technology, Atlanta, GA, USA
e-mail: cemasse@sandia.gov; cmassey36@gatech.edu

D. G. Moore
Sandia National Laboratories, Albuquerque, NM, USA
e-mail: dgmoore@sandia.gov

C. J. Saldana
George W. Woodruff School of Mechanical Engineering, Georgia Institute of Technology, Atlanta, GA, USA
e-mail: christopher.saldana@me.gatech.edu

A. Amirkhizi et al. (eds.), *Challenges in Mechanics of Time Dependent Materials, Mechanics of Biological Systems and Materials & Micro-and Nanomechanics, Volume 2*, Conference Proceedings of the Society for Experimental Mechanics Series,
https://doi.org/10.1007/978-3-030-86737-9_7

potentially causing high stress concentrations [4]. For the purposes of this paper, the authors will refer to lack of fusion porosity as irregular porosity. Automated labeling of individual pores as spherical or irregular porosity could aid in the operator's determination of whether a part is fit for use. Machine learning methods such as Mask Regional Convolutional Networks (Mask R-CNNs) can identify shapes in an image and draw a bounding box and a mask around their region of occurrence [5]. During training, the Mask R-CNN determines its features and adjusts the model accordingly. Machine learning methods such as support vector machines and k-nearest neighbors clustering can be employed by feeding in features for which to train. Machine vision techniques can be employed for edge detection, including Canny edge filters and gradient methods [6].

7.2 Methods

AlSi10Mg dogbone boats were printed with three conditions: high, nominal, and low global energy density (GED) values. The high and nominal GED conditions were found to have mostly spherical pores, whereas the low GED condition was found to have both spherical and irregular pores. These results were confirmed with SEM images. VGStudio Max 3.4 was used reconstruct CT images scanned using the Nikon M2 Dual Head 225/450 kV equipped with a Perkin Elmer detector. An ISO50 threshold was applied and the dogbone gauge image stacks were exported into .jpeg format for analysis in the *xy*, *xz*, and *yz* planes.

In total, 541 training images and 41 validation images were labeled with individual instances of porosity, as well as the type of porosity observed (i.e., spherical or irregular). All three GED conditions were represented in the sample testing and training sets. Image augmentation methods were applied to each image via horizontal flipping and vertical flipping, as well as either a 90° or 270° rotation. The pores were labeled using the VGG Image Annotator [7]. The authors used the Matterport Mask R-CNN architecture with the ResNet101 backbone [8]. Transfer learning methods were employed using the Microsoft COCO pretrained weights [9]. After many unsuccessful trial runs, in looking at the activation map, it was difficult to see the pores themselves. The activation map is responsible for showing any peaks or hot spots that may correspond with morphological features of objects in each class [10]. From the images in Fig. 7.1a–d, the noise of the grayscale values made it difficult to determine regions of porosity through multiple activation maps. Figure 7.1e–h shows interior pores as well as the edges being identified by the activation map. For this reason, the images had their pixels normalized by the average pixel of all the images in the set.

The authors conducted another experiment using a combination of machine learning and machine vision methods. A Canny edge detector was used in the Open-CV architecture to determine the edges of the porosity in the normalized images [6, 11]. Since the inputted images have some dead pixels and noise within the image, an algorithm was created to only

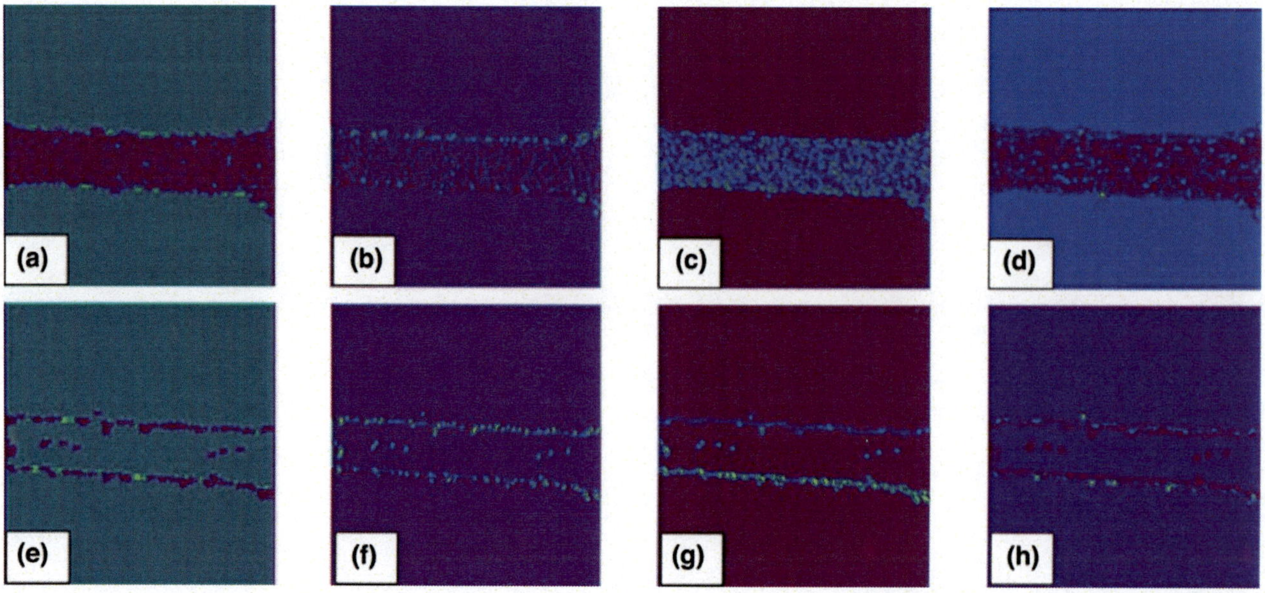

Fig. 7.1 **(a–d)** Sample activation map before normalization and **(e–h)** sample activation map after normalization

consider objects in the bounding box of the area of the edges of the dogbone. Metrics such as the x and y size of the pore, as well as the pore area and pore perimeter were inputted into the k-nearest neighbors and support vector machine algorithms using the Scikit-learn framework [12]. A sample set of labeled 80 images, including 531 instances of spherical porosity, 212 instances of non-porosity, and 153 lack of fusion porosity were used in this experiment.

7.3 Results and Discussion

Training time of the Mask R-CNN run was approximately 18 h using the NIVIDIA Titan GPU card using two images per GPU. The algorithm was extremely inaccurate and unable to determine porosity types. As shown in Fig. 7.2, the algorithm biased the edges in its determination of what it concluded as a pore. On instances where the algorithm identified internal porosity, it caught only a few pixels of the pore. It is possible that the pores were too small to work well with the scaling sliding window of the Mask R-CNN. The irregular-shaped pores morphology created some difficulty in being detected by the algorithm. More images could be used in future tests in addition to cropping and enlarging the size of the pores.

In the combination of machine learning and machine vision approach, the Canny edge filter segmented the porosity quite well. The results of the classification metrics of the KNN and SVM models are shown in Tables 7.1 and 7.2 below. The Linear Support Vector Machine had an overall accuracy of 0.78 compared to 0.81 for the KNN method. The recall metrics for the non-pores class were low for both the KNN and SVM methods, at 0.51 and 0.34, respectively. The inclusion of additional features, such as distance to the edge of the surface, may further help the algorithm differentiate these pores. Example classifications are shown in Fig. 7.3 and Table 7.3 below. From Fig. 7.3 and Table 7.3, both the SVM and KNN algorithms predicted the irregular and spherical porosity well, but determining the instances of non-porosity (in this case noise on the edge) was inaccurate and/or inconsistent for both the KNN and SVM. Holistically, the performance of the KNN and SVM are quite similar, although the KNN performs slightly better when considering the recall scores by class.

Fig. 7.2 Sample image from mask R-CNN implementation of porosity determination

Table 7.1 KNN classification metrics

	Precision	Recall	F1-score	Number of supporting elements
Irregular porosity	0.82	0.87	0.84	46
Non-pores	0.73	0.51	0.60	71
Spherical porosity	0.82	0.91	0.86	179
Macro average	0.79	0.76	0.77	296
Weighted average	0.80	0.81	0.80	296

Table 7.2 SVM classification metrics

	Precision	Recall	F1-score	Number of supporting elements
Irregular porosity	0.89	0.85	0.87	46
Non-pores	0.75	0.34	0.47	71
Spherical porosity	0.77	0.94	0.85	179
Macro average	0.80	0.76	0.73	296
Weighted average	0.78	0.78	0.76	296

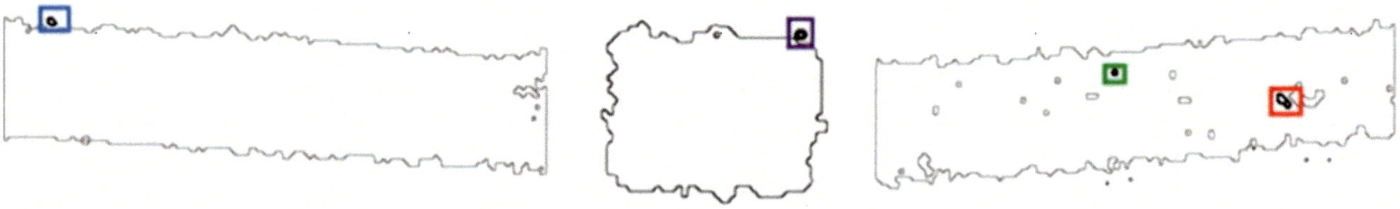

Fig. 7.3 Sample detected pores from combination approach with relevant information in Table 7.3 below

Table 7.3 Porosity predictions

	Red pore	Green pore	Blue pore	Purple pore
Ground truth	Irregular porosity	Spherical porosity	Non-pore	Non-pore
KNN prediction	Irregular porosity	Spherical porosity	Non-pore	Spherical porosity
SVM prediction	Irregular porosity	Spherical porosity	Spherical pore	Spherical porosity

7.4 Conclusion

The combination approach of machine vision and machine learning was more accurate in determining what class pores belonged to, as well as in segmenting porosity when compared to the Mask R-CNN approach. Additional image data of various image sizes could be used to further validate results. Future work could include adding additional features to help differentiate the items detected using the Canny edge detector that do not belong to either the spherical or irregular porosity class. A k-folds cross validation could be implemented in the code to further validate results. Other machine learning algorithms could be considered for the objective of improving precision and accuracy metrics.

Acknowledgments The authors would like to thank Sandia National Laboratories Machine Learning and Artificial Intelligence Group for their advice on this project. Special thanks Edward Jimenez, Andrew Lentfer, and Collin Epstein for their support in helping set up the computational architecture. The authors would also like to thank the National Defense Science and Engineering Graduate (NDSEG) Fellowship for their support. This work was also partially supported by DE-EE0008303 and NSF CMMI-1825640.

This paper describes objective technical results and analysis. Any subjective views or opinions that might be expressed in the paper do not necessarily represent the views of the U.S. Department of Energy or the United States Government.

Sandia National Laboratories is a multimission laboratory managed and operated by National Technology & Engineering Solutions of Sandia, LLC, a wholly owned subsidiary of Honeywell International Inc., for the U.S. Department of Energy's National Nuclear Security Administration under contract DE-NA0003525.

References

1. Sola, A., Nouri, A.: Microstructural porosity in additive manufacturing: the formation and detection of pores in metal parts fabricated by powder bed fusion. J. Adv. Manuf. Process. **1**(3) (2019)
2. du Plessis, A.: Effects of process parameters on porosity in laser powder bed fusion revealed by X-ray tomography. Addit. Manuf. **30** (2019)
3. Cunningham, R., Narra, S.P., Montgomery, C., Beuth, J., Rollett, A.D.: Synchrotron-based X-ray microtomography characterization of the effect of processing variables on porosity formation in laser power-bed additive manufacturing of Ti-6Al-4V. JOM. **69**, 479–484 (2017)
4. Kim, F.H., Moylan, S.P., Garboczi, E.J., Slotwinski, J.A.: Investigation of pore structure in cobalt chrome additively manufactured parts using X-ray computed tomography and three-dimensional image analysis. Addit. Manuf. **17**(2017), 23–28 (2017)
5. He, K., Gkioxari, G., Dollar, P., Girshick, R.: Mask R-CNN. In: Proceedings of the IEEE Conference on Computer Vision, pp. 2961–2969 (2017)
6. Canny, J.: A computational approach to edge detection. IEEE Trans. Pattern Anal. Mach. Intell. **PAMI-8**(6) (1986)
7. Dutta, A., Zisserman, A.: The via annotation software for images, audio and video. In: Proceedings of the 27th ACM International Conference on Multimedia, pp. 2276–2279 (2019)
8. Abdulla, W.: Mask R-CNN for Object Detection and Instance Segmentation on Keras and TensorFlow. GitHub Repository (2017)
9. Lin, T.Y., Maire, M., Belongie, S., Hays, J., Perona, P., Ramanan, D., Dollar, P., Zitnick, C.L.: Microsoft COCO: common objects in context. Computer vision-ECCV 2014. Lect. Notes Comput. Sci. **8693**, 740–755 (2014)
10. Zhu, Y., Zhou, Y., Xu, H., Ye, Q., Doermann, D., Jiao, J.: Learning instance activation maps for weakly supervised instance segmentation. In: Proceedings of the IEEE/CVF Conference on Computer Vision and Pattern Recognition, pp. 3116–3125 (2019)
11. Bradski, G.: The OpenCV library. Dr. Dobb's J. Softw. Tools. (2000)
12. Pedregosa, F., Varoquaux, G., Gramfort, A., Michel, V., Thirion, B., Grisel, O., Blondel, M., Prettenhofer, P., Weiss, R., Dubourg, V., Vanderplas, J., Passos, A., Cournapeau, D., Brucher, M., Perrot, M., Duchesnay, E.: Scikit-learn: machine learning in python. J. Mach. Learn. Res. **12**, 2825–2830 (2011)

Chapter 8
Constitutive Modeling of the Dynamic Behavior of Cork Material

Marco Sasso, Attilio Lattanzi, Emanuele Farotti, Fabrizio Sarasini, Claudia Sergi, Jacopo Tirillò, and Edoardo Mancini

Abstract In this work, compression tests have been carried out at different strain rate, from 10^{-3} to approximately 10^3 s^{-1}, on agglomerated cork material. The quasi-static and low strain rate tests have been conducted by means of servo-pneumatic machine, whereas the high strain rate tests have been conducted by means of polymeric Hopkinson bar. The experimental results show a stress–strain relationship that is characterized by a typical S-shaped curve. As expected, the strength is observed to increase when the material is deformed at increasing strain rate. In addition, the properties during the relaxation phase have been considered as well, showing that the stress response is characterized by a rapid decrement while the deformation is almost completely recovered. The global mechanical behavior is found to be very well reproduced by a combination of constitutive models, which include compressible hyperelastic modeling and large-strain viscoelasticity. The matching between the experimental and analytical data is very precise in the monotonic loading phase. Moreover, considering a damage model of the Mullins type it is possible to reproduce reasonably well also the unloading phase.

Key words Cork · Constitutive modeling · Viscoelasticity · Hyperelasticity · High strain rate

8.1 Introduction

Cellular materials are intensively used in engineering applications when lightweight and energy absorption capabilities are desired. Nowadays, particular attention is also given to environmental impact, which is related to the material processing, to its use during operative life, and to its disposal or reuse at the end of its life cycle. For this reason, natural materials are gaining attention in recent years [1–3]. Among cellular natural materials, cork is probably the highest spot [4]. Cork is natural, as it is obtained by the bark of Quercus suber tree and it is characterized by the major advantage of being biodegradable. A tree can be harvested approximately every 10 years, and it lasts for 100–150 years, hence cork is renewable. Moreover, it can be recycled for multiple uses [5].

Cork presents the typical mechanical behavior of a cellular material. Under compression, the stress–strain curve is characterized by an initial elastic region, followed by a plateau region where cell walls buckling occurs and most part of the deformation is accumulated; then, densification takes place and stress rapidly increases. The curve is then similar to the one that can be observed in foams, which are often adopted for energy dissipation purposes [6–8].

However, not only cork is capable of absorbing energy, but it also has a high recovery capacity. This means that the material deformation is mainly elastic, so that cork returns to its initial shape after load removal [9]. This can be very useful since the material keeps the energy absorption capacity almost constant, even after multiple impacts.

In this work, compression tests have been carried out at different strain rate, from 10^{-3} to approximately 10^3 s^{-1}, measuring the stress–strain behavior of the material. Not only the monotonic loading phase was recorded, but also the

M. Sasso (✉) · A. Lattanzi · E. Farotti
Dipartimento di Ingegneria Industriale e Scienze Matematiche, Università Politecnica delle Marche, Ancona, Italy
e-mail: m.sasso@univpm.it; a.lattanzi@pm.univpm.it; e.farotti@pm.univm.it

F. Sarasini · C. Sergi · J. Tirillò
Department of Chemical Engineering Materials Environment, Sapienza Università di Roma, Rome, Italy
e-mail: fabrizio.sarasini@uniroma1.it; claudia.sergi@uniroma1.it; jacopo.tirillo@uniroma1.it

E. Mancini
Department of Industrial and Information Engineering and Economics, Università degli studi dell'Aquila, l'Aquila, Italy
e-mail: edoardo.mancini@univaq.it

A. Amirkhizi et al. (eds.), *Challenges in Mechanics of Time Dependent Materials, Mechanics of Biological Systems and Materials & Micro-and Nanomechanics, Volume 2*, Conference Proceedings of the Society for Experimental Mechanics Series,
https://doi.org/10.1007/978-3-030-86737-9_8

unloading phase was considered, in order to evaluate the possible damage evolution experienced by the material and the elastic energy release that may occur after an impact. Successively, the experimental curves have been used to calibrate a constitutive model, which includes the main features of the mechanical response of cork, i.e., elasticity, viscoelasticity, and damage.

8.2 Materials and Methods

The specimens tested in this work are rectangular blocks of agglomerated cork (Fig. 8.1a), approximately cubic in shape with a side of 10 mm and with a density of approximately 140 kg/m^3. The samples have been tested in compression in the strain rate range from 10^{-3} to 10^3 s^{-1} and the tests at 10^{-3}, 10^{-1} and 10^1 s^{-1} were conducted by means of a pneumatic testing machine, model Siplan® (Fig. 8.1b), equipped with a 3 kN load cell and capable of reaching 100 mm/s piston speed.

For the tests at 10^{-3} and 10^{-1} s^{-1}, the samples were simply placed onto a flat plate connected to the load cell, on the fixed crosshead of the machine and the piston was gently moved to approach the upper specimen surface to start the test. For the tests at 10^1 s^{-1}, the piston was moved upwards, far enough from the sample to complete the acceleration ramp from 0 to 100 mm/s. In this way, the samples were deformed at the most constant speed possible with the used machine. To evaluate the relaxation phase of the material in the tests at 10^{-3} and 10^{-1} s^{-1}, the deformation was maintained constant at the end of the loading ramp for an amount of time equal to the ramp duration; then, the load is removed by moving the piston back, until the load is zeroed. The motion law assigned to the piston is represented by the trapezoid shape illustrated in Fig. 8.1c. In the test at 10^1 s^{-1}, instead, the piston has been assigned a triangular waveform, where the unloading ramp starts immediately after the loading ramp, as shown by the dashed red line in Fig. 8.1c.

The dynamic tests at 10^3 s^{-1} have been performed by means of the Split Hopkinson Pressure Bar (SHPB) shown in Fig. 8.2, whose calibration and adaptation to test soft materials are described in [10–12]. In addition, the SHPB setup was such that the specimen relaxation was partially evaluated, by extending the measurement of the reflected and transmitted waves. It must be highlighted that in this kind of tests, the tool motion that deforms the sample cannot be exactly prescribed as in Fig. 8.1c; however, the obtained displacement history was found to be very similar to a triangular waveform. Hence, it was possible to measure partially the unloading phase after the maximum load, with no holding phase.

The imposed strain histories for the four tests are shown in Fig. 8.3.

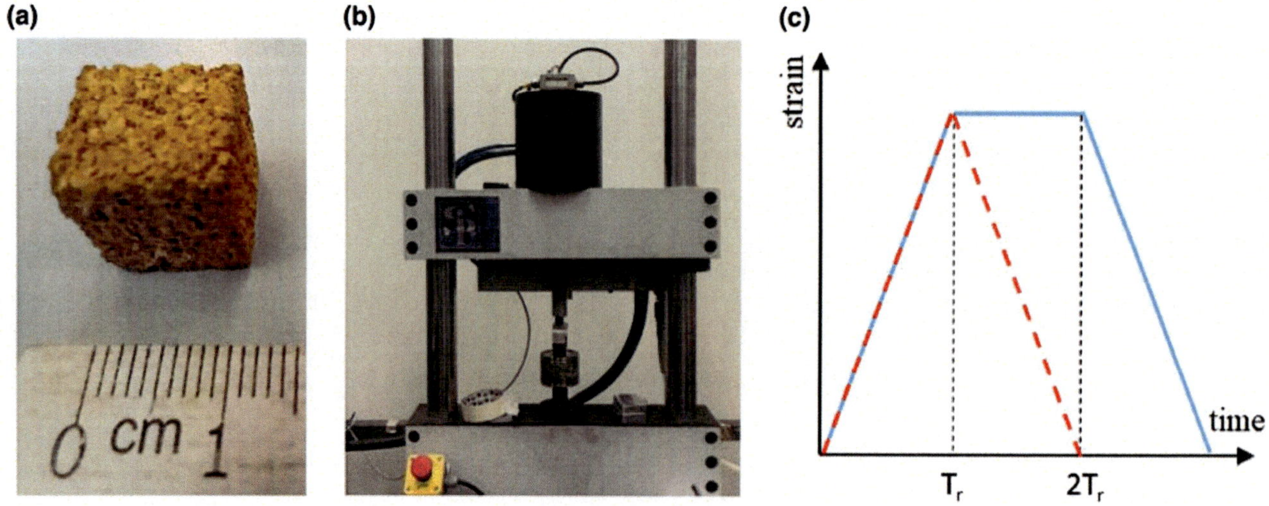

Fig. 8.1 (**a**) Agglomerated cork sample, (**b**) servo-pneumatic testing machine, (**c**) imposed strain histories

Fig. 8.2 Split Hopkinson pressure bar installed at Marche Polytechnic University

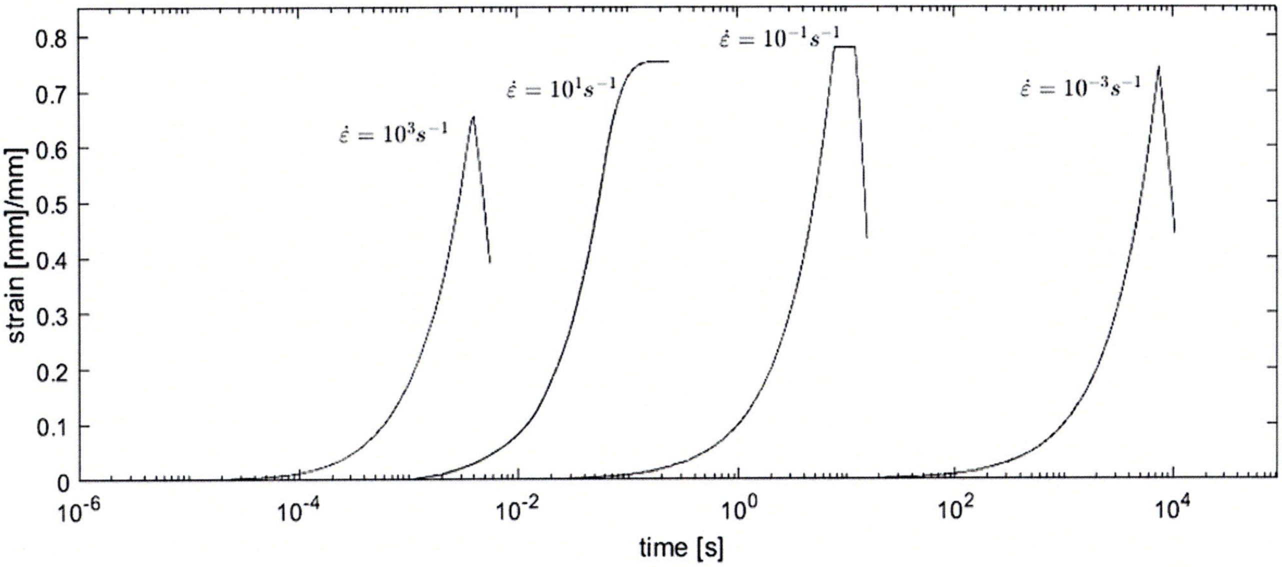

Fig. 8.3 Strain histories imposed in the compression tests at different strain rates

8.3 Material Modeling

The constitutive modeling approach used attempts to describe the principal mechanical properties of cork when subjected to dynamic loading. As reported in the literature [13], the S-shaped stress–strain curve obtained in the quasi-static test suggests that the material can be modeled by a hyperelastic law, where the stress is in relationship with a strain energy potential W. The experimental evidence also showed that, as well known, the cork exhibits an almost perfectly null Poisson's ratio, meaning that a compressible hyperelastic model is best suited. Among the several models available in the literature, the Ogden Compressible Foam model, which is implemented as a built-in material within the Abaqus FE commercial code as "Hyperfoam," was adopted in this work. The strain energy function of order N is defined:

$$W = \sum_{i=1}^{N} \frac{2\mu_i}{\alpha_i^2} \left[\lambda_1^{\alpha_i} + \lambda_2^{\alpha_i} + \lambda_3^{\alpha_i} - 3 + \frac{1}{\beta_i} \left(J^{-\alpha_i \beta_i} \right) \right] \tag{8.1}$$

where α_i and β_i and μ_i are material constants, λ_{1-3} are the stretch ratios in the three principal directions, and J is the volume ratio, $J = \lambda_1 \lambda_2 \lambda_3$. The principal stresses are computed taking the partial derivative of Eq. (8.1) with respect to λ. The null Poisson's ratio determines that $\beta_i = 0$; in uniaxial compression state $\lambda_1 = \lambda_2 = 1$, and the nominal stress becomes:

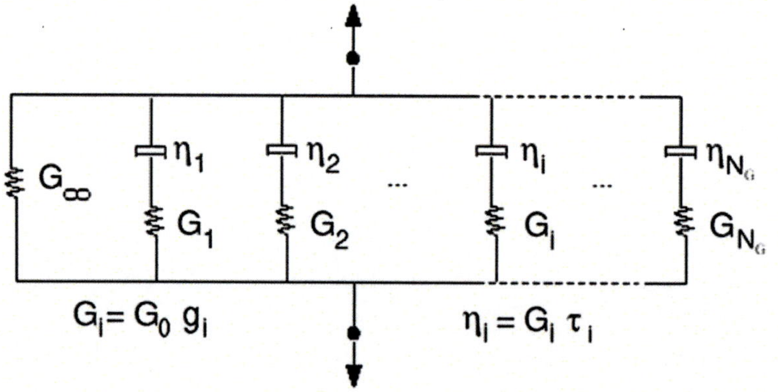

Fig. 8.4 Generalized Maxwell model used to represent the viscoelastic behavior (Prony series)

$$\sigma_3 = \frac{\partial W}{\partial \lambda_3} = \sum_{i=1}^{N} \frac{2\mu_i}{\alpha_i \lambda_3}\left(\lambda_3^{\alpha_i} - 1\right) \tag{8.2}$$

Equation (8.2) can be used to fit the experimental stress–strain curve in quasi-static condition, identifying the best μ_i and α_i coefficients. Cork material is well known to be viscoelastic [14], indeed its mechanical response strongly depends on the strain rate. In view of this, the viscoelastic behavior has to be modeled as well, in conjunction with the Hyperfoam constitutive description. The Prony series has been used in this work, which consists in considering the material as a parallel combination of one hyperelastic spring, which represents the long-term stiffness G_∞, with a given number, N_G, of hyperelastic Maxwell branches, as shown in Fig. 8.4. Each Maxwell layer is characterized by a relaxation time τ_i. The global relaxation module of such a system is given by:

$$G(t) = G_0\left[g_\infty + \sum_{i=1}^{N_G}\left(g_i e^{-t/\tau_i}\right)\right] \tag{8.3}$$

The framework used is the large-strain viscoelasticity, which also means that the N_G springs are non-linear, hyperelastic, springs. In Eq. (8.3), g_i are the relative moduli, so that $(g_\infty + \sum g_i) = 1$; G_0 represents the short term, or instantaneous, stiffness that is obtained as the sum of all branches' stiffness. The long-term and short-term stiffnesses are related as $G_\infty = (g_\infty \cdot G_0)$. The stress along time in the i-th layer of the Generalized Maxwell model is computed exploiting the push-forward Simo Scheme [15], which leads to:

$$\sigma_i(t + \Delta t) = a_i g_i \sigma_o(t + \Delta t) + b_i g_i \sigma_o(t) + c_i \sigma_i(t) \tag{8.4}$$

with $c_i = \exp(-\Delta t/\tau_i)$, $b_i = (1 - c_i)\tau_i/\Delta t - c_i$ and $a_i = 1 - (1 - c_i)\tau_i/\Delta t$. In Eq. (8.4), σ_0 is the stress corresponding to the purely hyperelastic behavior, considering the instantaneous moduli. The total stress becomes:

$$\sigma(t + \Delta t) = \sigma_0(t + \Delta t) - \sum_{i=1}^{N_G} \sigma_i(t + \Delta t) \tag{8.5}$$

Finally, in order to describe the material behavior also in the unloading phase, the Mullins damage model has been adopted. The model is well established for rubberlike materials [16], but it has been adopted for cork material as well [17]. It consists in the introduction of a damage variable η, which is used to reduce the stress with respect to the hyperelastic stress of the virgin material:

$$\sigma(\eta, \lambda) = \eta\sigma(\lambda) \tag{8.6}$$

Equation (8.6) is used when the material is at an energy potential W that is lower than the maximum energy potential W_m experienced by the material itself at the end of the loading phase. It means that η is equal to 1 along the primary loading path, while it is between 0 and 1 during the unloading path. The η term is generally assumed to be represented by the so called "error function":

$$\eta = 1 - \frac{1}{r}\,\mathrm{erf}\left(\frac{W_m - W}{m + \beta W_m}\right) \tag{8.7}$$

where r, m and β are material parameters that govern the shape of the unloading curve. Note that $r > 1$, $m \geq 0$, $\beta > 0$. Since the unloading curve is lower than the primary loading one, a dissipative cycle is established which contributes, together with the viscoelastic hysteresis cycle, to the global energy dissipation of cork.

8.4 Results

The models coefficients have been computed by solving an optimization problem, where a cost function represented by the difference between the experimental and analytical stresses is minimized. In the analytical model, the experimentally measured strain and strain rate are introduced into Eqs. (8.2), (8.4) and (8.6). A fourth order Hyperfoam model was used for describing the compressible hyperelastic behavior meaning that 8 parameters have to be found, i.e., μ_i and α_i with $i = 1 \ldots 4$. A 10 terms Prony series were used for describing the viscoelastic behavior, where the Maxwell branches have equally spaced relaxation times in the logarithmic scale meaning that the minimization problem aims at determining the 2 extreme relaxation times $\tau_{\min}$, $\tau_{\max}$, and 10 relative moduli g_i. Finally, the Mullins damage model requires finding further 3 parameters, bringing the total to 23 coefficients. The calibration has been implemented within a Matlab® optimization procedure.

The fitting coefficients of the Hyperfoam model (long term) are reported in Table 8.1. The best set of Prony series coefficients are reported in Table 8.2.

Analyzing the relative moduli, it appears that their trend is not monotonic with respect to the corresponding relaxation times. Indeed, g_4, g_5, g_8 and g_9, which correspond to relaxation times $6 \cdot 10^{-4}$, $4.9 \cdot 10^{-3}$, $2.4 \cdot 10^{0}$, $1.9 \cdot 10^{1}$ s, are found to be very close to zero, while the other Maxwell layers, especially g_1, g_2 and g_3 that correspond to $1.2 \cdot 10^{-6}$, $1.0 \cdot 10^{-5}$, and $7.9 \cdot 10^{-5}$ s, are found to be much larger. Table 8.3 shows the coefficients of the Mullins damage model that best fit the unloading phase. The combination of small values for both m and r, close or at the lower limit of the admissible range, determine a very large dissipation cycle, characterized by a sudden stress drop just at the beginning of the unloading phase.

The results of the experimental tests and of the analytical modeling, in terms of stress–strain curves, are reported in Fig. 8.5. The black dots represent the experimental points, the solid line in red represents the long-term, or purely hyperelastic, numerical stress, the dashed lines represent the stress in the different layers of the Generalized Maxwell model (only the loading phase is reported for the sake of clarity), and the solid line in green represents the total stress predicted by the constitutive model. It is observed that the model is able to describe very well the material behavior in all phases, including the primary loading path, the holding phase, and the unloading phase. The global root mean square deviation between numerical and experimental data is 0.11 MPa.

Table 8.1 Fourth order hyperfoam coefficients

μ_1	μ_2	μ_3	μ_4	α_1	α_2	α_3	α_4	$\beta_{1\ldots4}$
1.13	1.45	0.04	−1.72	4.30	3.16	−1.79	2.73	0

Table 8.2 Prony series coefficients

$\tau_{\min}$	$\tau_{\max}$	g_1	g_2	g_3	g_4	g_5	g_6	g_7	g_8	g_9	g_{10}	g_∞
E − 5.98	E + 2.10	0.074	0.404	0.384	0.000	0.008	0.022	0.039	0.000	0.001	0.010	0.056

Table 8.3 Mullins damage coefficients

r	m	β
1.01	0	0.49

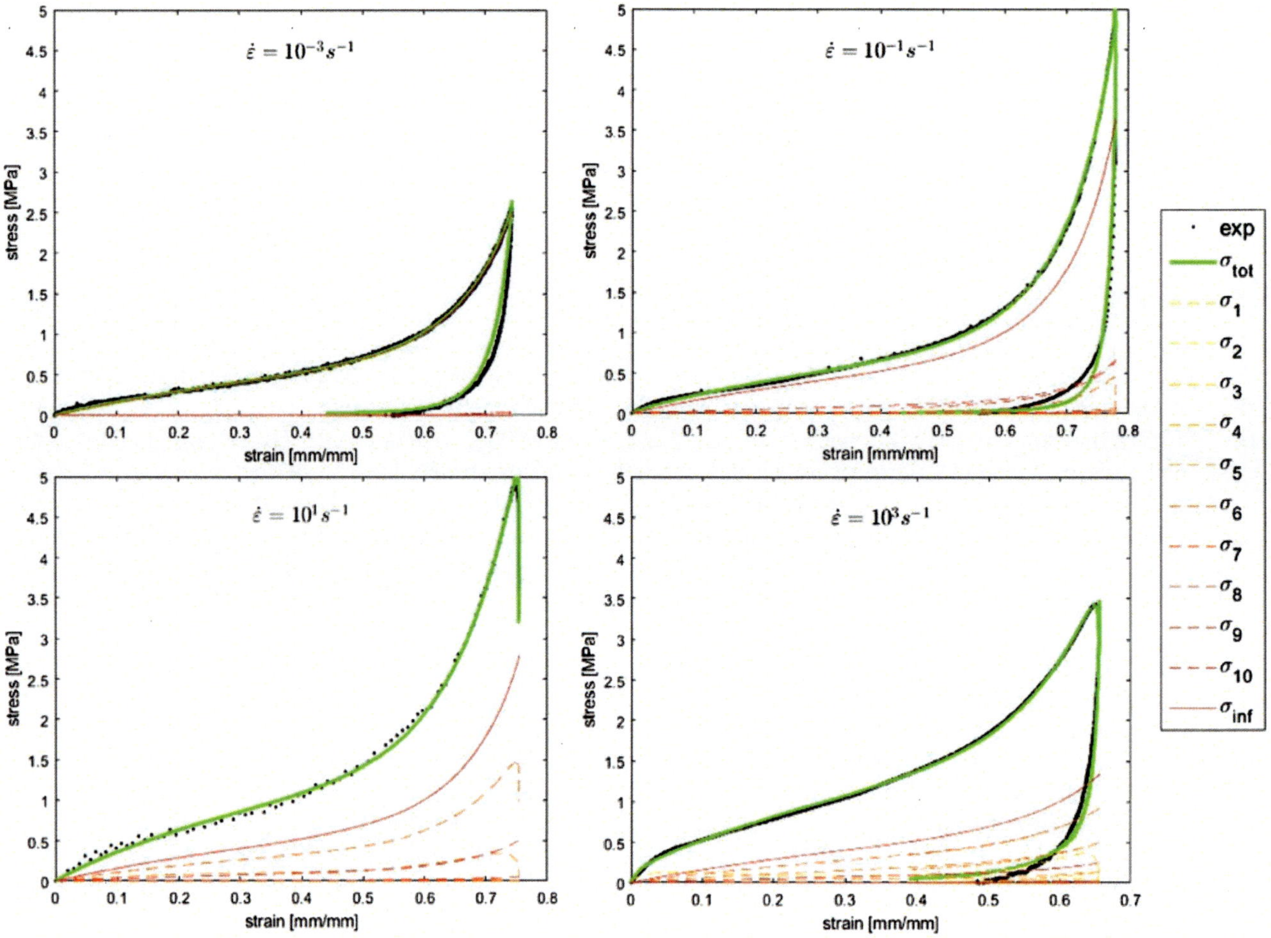

Fig. 8.5 Experimental and numerical stress–strain curves

8.5 Conclusion

The work proposed the mechanical characterization and constitutive modeling of an agglomerated cork subjected to compression loading. The stress–strain curve is characterized by the typical S-shaped trend, which is very similar to that of synthetic foam materials, and is found to be very well represented by a compressible hyperelastic formulation. The tests conducted at different strain rates highlighted the time dependency of the stress–strain relationship, hence a 10 term Prony series has been used to describe the strength increase in dynamic conditions. In addition, in the experimental tests, the primary monotonic loading phase has been followed by holding and unloading phases and this permitted to evaluate the damage behavior of cork, which is found to be well represented by the Mullins model. The three constitutive models can be made to coexist in order to give a complete description of the cork behavior when it is subjected to dynamic loads and impacts.

References

1. Castro, O., Silva, J.M., Devezas, T., Silva, A., Gil, L.: Cork agglomerates as an ideal core material in lightweight structures. Mater. Des. **31**, 425–432 (2010)
2. Sarasini, F., Tirillò, J., Lampani, L., Sasso, M., Mancini, E., Burgstaller, C., Calzolari, A.: Static and dynamic characterization of agglomerated cork and related sandwich structures. Compos. Struct. **212**, 439–451 (2019)
3. Sanchez-Saez, S., García-Castillo, S.K., Barbero, E., Cirne, J.: Dynamic crushing behavior of agglomerated cork. Mater. Des. **65**, 743–748 (2015)

4. Silva, S.P., Sabino, M.A., Fernandes, E.M., Correlo, V.M., Boesel, L.F., Reis, R.L.: Cork: properties, capabilities and applications. Int. Mater. Rev. **50**, 345–365 (2005)
5. Pereira, H.: Cork: Biology, Production and Uses. Elsevier (2007)
6. Gibson, L.J., Ashby, M.F.: Cellular Solids: Structure and Properties, vol. 7. Cambridge University Press (1999)
7. Koohbor, B., Kidane, A., Wei-Yang Lu, L.: Characterizing the constitutive response and energy absorption of rigid polymeric foams subjected to intermediate-velocity impact. Polym. Test. **54**, 48–58 (2016)
8. Fernandes, F.A.O., Jardin, R.T., Pereira, A.B., Alves de Sousa, R.J.: Comparing the mechanical performance of synthetic and natural cellular materials. Mater. Des. **82**, 335–341 (2015)
9. Sergi, C., Sarasini, F., Tirillò, J., Barbero, E., Sanchez-Saez, S., Sasso, M., Mancini, E.: Temperature, strain rate and anisotropy effects on compressive response of natural and synthetic cellular core materials. Compos. Struct. **260**, 113268 (2015). https://doi.org/10.1016/j.compstruct.2020.113268
10. Sasso, M., Antonelli, M., Mancini, E., Radoni, M., Amodio, D.: Experimental and numerical characterization of a polymeric Hopkinson bar by DTMA. Int. J. Impact Eng. **103**, 50–63 (2017)
11. Martarelli, M., Mancini, E., Lonzi, B., Sasso, M.: Sensor calibration of polymeric Hopkinson bars for dynamic testing of soft materials. Meas. Sci. Technol. **29**(2), 025601 (2018)
12. Farotti, E., Mancini, E., Bellezze, T., Sasso, M.: Investigation of the effects of mold temperature, test temperature and strain rate on mechanical behaviour of polypropylene. J. Dynam. Behav. Mater. **5**, 344–360 (2019)
13. Fernandes, F.A.O., Pascoal, R.J.S., Alves de Sousa, R.J.: Modelling impact response of agglomerated cork. Mater. Des. **58**, 499–507 (2014)
14. Mano, J.F.: The viscoelastic properties of cork. J. Mater. Sci. **37**, 257–263 (2002)
15. Simo, J.C.: On a fully three-dimensional finite-strain viscoelastic damage model: formulation and computational aspects. Comput. Methods Appl. Mech. Eng. **60**, 153–173 (1987)
16. Sasso, M., Amodio, D.: Development of a biaxial stretching machine for rubbers by optical methods. In: Proceedings of the 2006 SEM Annual Conference and Exposition on Experimental and Applied Mechanics, vol. 3, pp. 1161–1171 (2006)
17. Ptak, M., Kaczynski, P., Fernandes, F.A.O., Alves de Sousa, R.J.: Assessing impact velocity and temperature effects on crashworthiness properties of cork material. Int. J. Impact Eng. **106**, 238–248 (2017)

Chapter 9
The Penetration Dynamics of a Violent Cavitation Bubble Through a Hydrogel–Water Interface

Jin Yang, Yue Yin, Harry C. Cramer III, and Christian Franck

Abstract Understanding the dynamical behavior of an oscillating bubble near a hydrogel–water interface is an interesting and important multiphase problem arising in many medical treatments including minimizing tissue damage during ultrasound and laser surgeries, guiding targeted drug deliveries, to name a few. Here, by using ultrahigh-speed videography, we captured the complex interaction of an inertial cavitated bubble at a soft hydrogel–water interface penetrating the gel-fluid (water) boundary. Next, we experimentally measured and numerically modeled the nonlinear bubble dynamics near the hydrogel–water interface. Here we present our experimentally observed interface penetration process including annular jetting and shock wave propagation toward the water side. On the gel side we observed the induction of significant large and complex deformations induced by the cavitation bubble. We provide a comprehensive analysis of these phenomena including a quantitative estimate of the associated material strains and damage during this high strain-rate penetration process.

Key words Inertial cavitation · High strain-rate · Interface · Penetration · Finite deformation

9.1 Introduction

Cavitation is a common phenomenon in many biological systems and medical applications [1, 2]. For example, inertial cavitation can cause serious damage to soft tissues and cells [3–5]. On the other side, when harnessed carefully, bubble cavitation can be used beneficially in many surgical and medical procedures, for example, in cataract laser surgery, lithotripsy and histotripsy applications [6, 7].

In the past three decades, the dynamics of a cavitating bubble along a boundary was found to have significant influence on the bubble dynamics itself. For example, the interaction of an inertial cavitation bubble with a nearby liquid–liquid interface is a complex multiphase problem that a liquid jet can be induced to drill into the heavier liquid during the bubble collapse [8]. For a laser-induced cavitation bubble near a gel–water interface, the deformation and rebound of the boundary and the Bjerknes attraction force can induce bubble splitting, liquid jets formation, and jet-like ejection of the boundary material into the liquid [9–11]. The jetting behavior was found to depend on the distance between the laser focal point and the interface, and the interface elastic modulus.

Here, by recording the spatiotemporally resolved bubble dynamics via high-speed videography, a combined experimental and theoretical investigation of single bubble cavitation near a gel–water interface has been conducted. Compared with previous studies, the gel we tested here is more compliant with a water concentration above 90% and an elastic modulus on the order of $O(0.1) \sim O(1)$ kPa. In addition to bubble splitting and annular jetting, we also observed the formation of a vortex ring at the first violent collapse. On the gel side, when bubble was close to the gel–water interface, we found gel-material was torn off and created a hole at the first collapse. We provide a comprehensive analysis of these phenomena including a quantitative

J. Yang · Y. Yin · C. Franck (✉)
Department of Mechanical Engineering, University of Wisconsin – Madison, Madison, WI, USA
e-mail: jyang526@wisc.edu; yyin49@wisc.edu; cfranck@wisc.edu

H. C. Cramer III
Center for Biomedical Engineering, Brown University, Providence, RI, USA

Department of Mechanical Engineering, University of Wisconsin – Madison, Madison, WI, USA

School of Engineering, Brown University, Providence, RI, USA
e-mail: harry_cramer@brown.edu

A. Amirkhizi et al. (eds.), *Challenges in Mechanics of Time Dependent Materials, Mechanics of Biological Systems and Materials & Micro-and Nanomechanics, Volume 2*, Conference Proceedings of the Society for Experimental Mechanics Series,
https://doi.org/10.1007/978-3-030-86737-9_9

estimate of the associated material strains and damage during this high strain-rate penetration process. These results might be of significant interest in medical and engineering applications with respect to cavitation erosion, collateral damage in laser surgery, cavitation-mediated enhancement of pulsed laser ablation of tissue, and guiding targeted drug deliveries.

In this paper, first, we introduce our material preparation and experimental setup in Sects. 9.2 and 9.3. Then we show and analyze our experimental results in Sect. 9.4. Finally, we present our conclusions in Sect. 9.5.

9.2 Material Preparation

In our experiments, soft polyacrylamide (PA) hydrogel samples are prepared from 40.0% acrylamide solution and 2.0% bis solution (Bio-Rad, Hercules, CA) mixed to a final concentration of 3.0%/0.2% Acrylamide/Bis (v/v) in deionized water and crosslinked with 0.5% APS (ThermoFisher Scientific, USA) and 1.25% TEMED (ThermoFisher Scientific, USA). Once mixed, PA samples are completely submersed in deionized water for 24 h to allow for complete swelling. The quasistatic shear modulus, G_{∞}, of the prepared soft PA is 0.57 kPa. Soft PA has also been found to show strain stiffening effects during our previous inertial cavitation experiments and is quantitatively characterized by a quadratic law Kelvin-Voigt (qKV) viscoelastic material model with a strain stiffening parameter $\alpha = 0.96 \pm 0.058$, and viscosity $\mu = 0.060 \pm 0.057$ Pa s [12].

9.3 Experimental Setup

The experimental setup in this study is adapted from our previous studies [12–15] as shown in Fig. 9.1, where single cavitation bubbles were generated through a single pulse from an adjustable 1–25 mJ Q-switched Nd:YAG Minilite II (Continuum, Milpitas, CA) laser platform frequency doubled to 532 nm [13]. Laser pulses were expanded to fill the back aperture of a Nikon Plan Fluor 20X/0.5 NA imaging objective and were aligned through the back camera port of a Nikon Ti:Eclipse microscope (Nikon Instruments, Long Island, NY). Pulses were reflected off a 532 nm notch dichroic mirror (Semrock, Rochester, NY) to the rear aperture of the imaging objective, and then were focused on the interface between the hydrogel and the water. Cavitation bubbles were recorded at one million fps with a Kirana5M high-speed camera (Specialized Imaging, Pitstone, United Kingdom), with triggered full-field illumination from a SILUX640 laser illumination system (Specialized Imaging, Pitstone, United Kingdom). Samples were imaged in 25 mm Chamlide magnetic chambers (Live Cell Instrument, Seoul, South Korea) in all tests. Output TTL signals from the camera aligned image acquisition with the laser pulse and illumination pulses. An exposure time of 500 ns and an illumination pulse width of 250 ns were used to maximize illumination

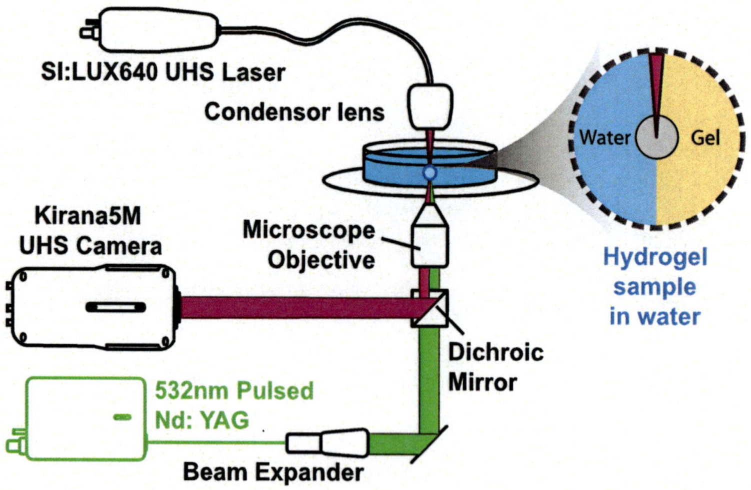

Fig. 9.1 Experimental setup of laser-induced inertial micro-cavitation. A single 6 ns, Q-switched 532 nm Nd:YAG laser pulse of 1–10 mJ passes through a beam expander to fill the back aperture of an objective mounted into an inverted TI-Eclipse microscope, and is focused on the interface of soft PA hydrogel and water. (Image modified based on Yang et al. [12])

and minimize image ghosting in all samples at 924 × 768 pixels for the full 180 acquisition frames. A Q-switch delay: 150 μs was used to reduce pulse to pulse energy variability. Cavitation events were generated 600 μm above the bottom surface to ensure boundary effects remained negligible.

9.4 Results and Discussions

By using ultrahigh-speed videography, we captured the complex interaction of an inertial cavitated bubble arising at the hydrogel-water interface. Each individual acquisition frame was analyzed to determine the effective bubble radius at each time-step using a custom-written MATLAB script. An example of fitted bubble radius vs. time curve is shown in Fig. 9.3a. The maximum bubble radius R_{max} is around 320 μm. The bubble was captured to undergo nine expansion-collapse cycles within 140 ~ 170 μs. The maximum radial stretch ratio is around 8.

Images of cavitated bubbles at selected time points are further shown in Fig. 9.2a–l, where all the time points are shifted by a constant so that the bubble radius reaches its maximum value at time zero. In each of the frames in Fig. 9.2a–f before the first collapse, a red dashed line is overlaid onto the original image to visualize the shape of the bubble. Similarly, after the first collapse, blue dashed lines are overlaid to visualize the gel–water interface in Fig. 9.2f–l. We found that the shape of the bubble maintains spherical symmetry during its first expansion (cf Fig. 9.2a, b). The bubble became non-spherical during the first collapse (cf Fig. 9.2d–f) because the hydrogel material was stiffer and interacted more strongly than water. During the first violent collapse, an annular jetting was developed toward the water side as shown in Fig. 9.2g, h, which was similar to the Bjerknes effect of inertial cavitation near an elastic solid [9–11]. A vortex ring was also formed, and shock waves were released into the water side (cf Fig. 9.2g). Unlike previous studies [9–11], we found that the gel–water interface was peeled off into the water. During the second expansion-collapse cycle, the bubble kept migrating toward the water side and split into

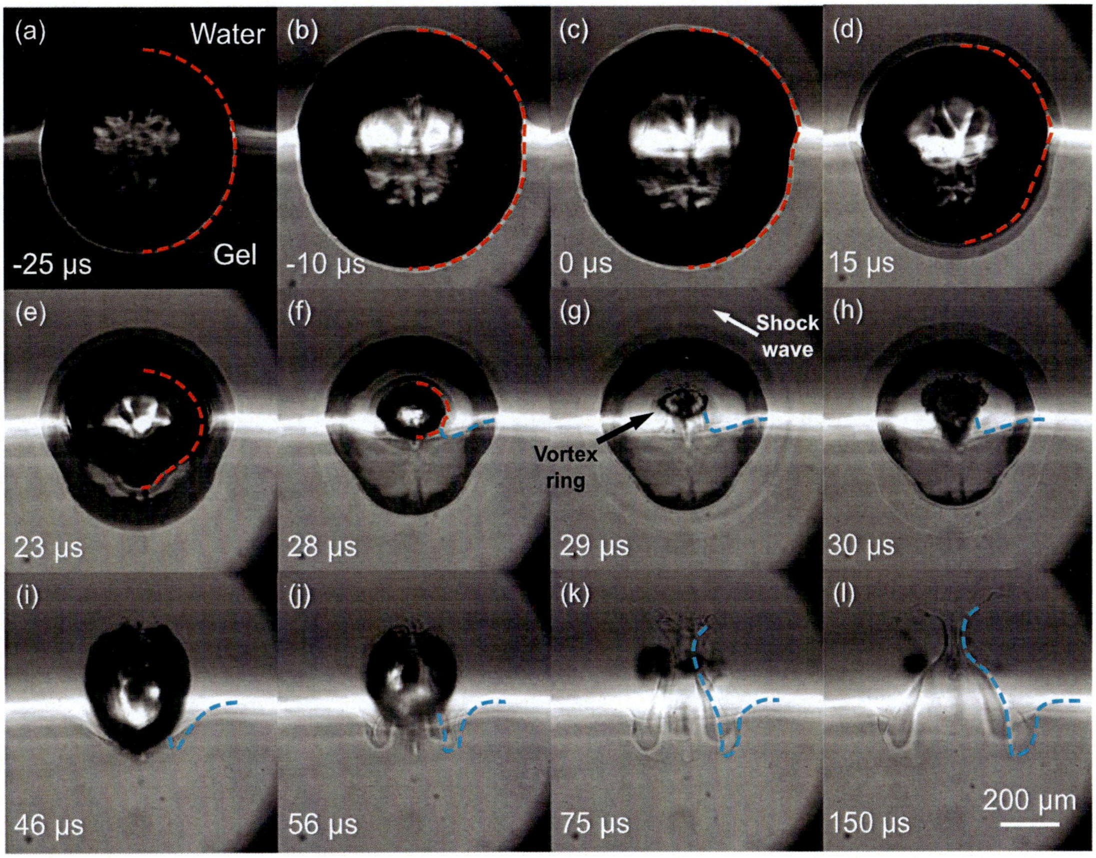

Fig. 9.2 Time-lapse images of a single cavitation bubble near a gel–water interface. Red dashed lines are overlaid onto the original image to visualize the shape of the bubble in (**a–f**). After the first collapse, blue dashed lines are overlaid to visualize the gel–water interface in (**f–l**). (Arrows in (**g**): during the first violent collapse, a vortex ring was formed, and shock waves were released)

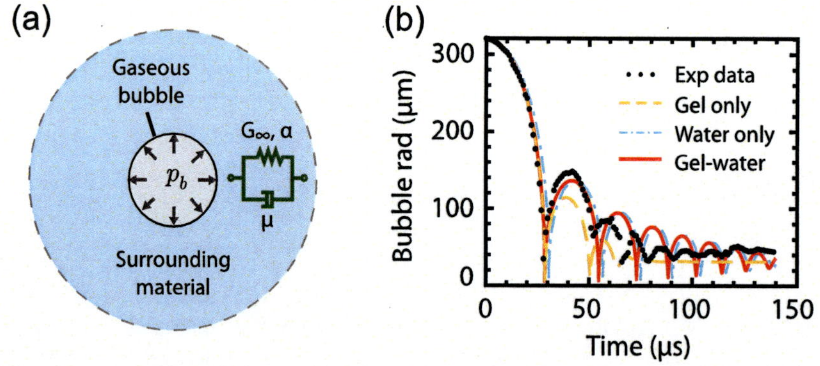

Fig. 9.3 (**a**) Theoretical model of inertial cavitation in a viscoelastic soft material. (**b**) Experimentally measured and numerically simulated bubble effective radius vs. time curves

smaller bubbles. The gel–water interface deformed into a dome shape and bent toward the water side finally breaking at its apex (cf Fig. 9.2i, j). Finally, the smaller bubbles dissolved in the water and the gel–water interface exhibited viscoelastic behavior, releasing its residual strain as shown in Fig. 9.2j–l.

To model the bubble dynamics, we apply the Keller-Miksis equation (9.1). Though the spherical symmetry is only well-maintained during bubble's first expansion process, we approximate that Eq. (9.1) still holds for the bubble's effective radius, i.e.,

$$\left(1-\frac{\dot{R}}{c}\right)R\ddot{R}+\frac{3}{2}\left(1-\frac{\dot{R}}{3c}\right)\dot{R}^2=\frac{1}{\rho}\left(1+\frac{\dot{R}}{c}+\frac{R}{c}\frac{d}{dt}\right)\left(p_b-p_\infty+S-\frac{2\gamma}{R}\right) \tag{9.1}$$

where $R(t)$ is the evolution of the bubble's effective radius, and overdots denote derivatives with respect to time; ρ is the mass density of the surrounding material; c is the longitudinal wave speed; γ is the surface tension between gaseous bubble contents and the surrounding material; p_b is the internal bubble pressure; p_∞ is the far-field pressure assumed to be atmospheric; S is the stress integral of the deviatoric Cauchy stress as shown in Eq. (9.2). We model the surrounding material as a uniform, isotropic gel or water, depending on whether the bubble centroid is located at the gel side or the water side, called "gel-water" model. We also simulate the bubble dynamics in the pure gel ($G_\infty = 0.57$ kPa; $\alpha = 0.96 \pm 0.058$; $\mu = 0.060 \pm 0.057$ Pa s) and pure water ($G_\infty = 0$ Pa; $\mu = 0.001$ Pa s) as two comparisons. Finally, all the three simulation results are compared with the experimentally measured effective bubble radius in Fig. 9.3b.

$$S=\frac{(3\alpha-1)G_\infty}{2}\left[5-\left(\frac{R_\infty}{R}\right)^4-\frac{4R_\infty}{R}\right]-\frac{4\mu\dot{R}}{R}+2\alpha G_\infty\left[\frac{27}{40}+\frac{1}{8}\left(\frac{R_\infty}{R}\right)^8+\frac{1}{5}\left(\frac{R_\infty}{R}\right)^5+\left(\frac{R_\infty}{R}\right)^2-\frac{2R}{R_\infty}\right] \tag{9.2}$$

From Fig. 9.3b, we find that although the bubble is located at the gel–water interface, the kinetics of the effective bubble radius before the first R_{min} is almost identical to that in a pure isotropic gel. After first violent collapse ($t > 28$ μs), bubble has migrated toward the water side and its dynamics is close to that in pure water. The proposed "gel-water" model makes good agreement with the experimental measurements during both the first and second buble expansion-collapse cycles. After the second collapse ($t > 50$ μs), the single bubble splits into smaller bubbles, and none of these three models fits the experimental data very well because there are more complex physics involved.

Next, we quantitatively estimate the associated material strains and damage during this high strain-rate penetration process. Several geometry points are defined on the dome-shaped gel–water interface as shown in Fig. 9.4a, b, where T is the dome top point, A and B are intersections between the dome and the far-field gel–water interface. The dashed line in Fig. 9.4b indicates where the gel–water interface was broken. After the second collapse, points A and B moved to the left and right, respectively, and point T moved down to the horizontal axis. Points C and D are the left and right crimp tips at the dome bottom. We found that after $t \sim 80$ μs, the arc length of $\widehat{ATB}$ is almost constant value 495.8 ± 85.7 μm, which is further assumed to be the length of line $\overline{AB}$ in the final stress-free state. We also assume that the finite deformation of the gel–water interface is *cylindrically symmetric*:

$$r = r_0 + u_r; \quad z = z_0 + u_z \tag{9.3}$$

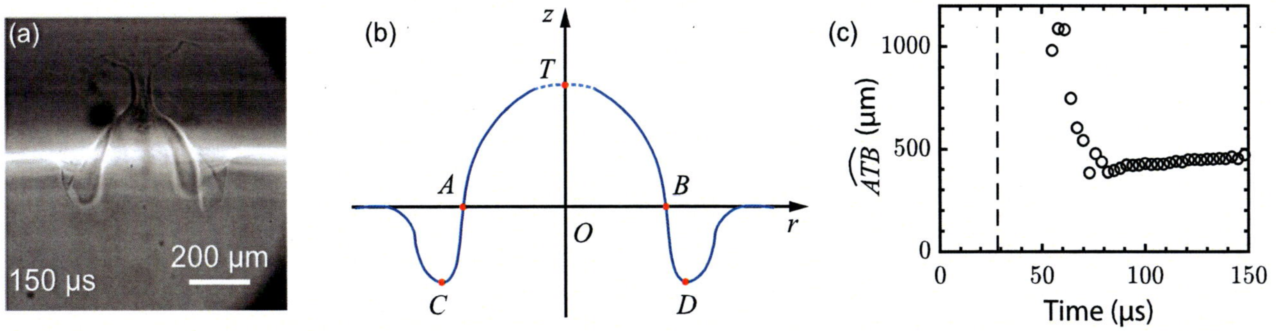

Fig. 9.4 (**a**) A typical frame of deformed, damaged gel–water interface. (**b**) Schematic of deformed gel–water interface, where the dashed curve near point T indicates that a hole was created at the apex of the dome. (**c**) Measured arc length $\widehat{ATB}$. (The vertical dashed line indicates when the first violent collapse happens)

where $\{r_0, z_0\}$ and $\{r, z\}$ are coordinates in the final stress-free and current deformed states; u_r and u_z are displacement components in the r and z directions, respectively. We further hypothesize that the deformation of the gel–water interface follows ansatz Eq. (9.4), where the current interface curve $\widehat{ATB}$ is approximated by a parabolic function:

$$\begin{cases} u_r(r, \theta, z_0 = 0) = (\beta - 1) r_0 \\ u_z(r, \theta, z_0 = 0) = \alpha_1 r^2 + \alpha_2 \end{cases} \tag{9.4}$$

where β is the radial compression ratio r/r_0; α_1, α_2 are two coefficients of the gel–water interface parabolic function. We also assume that soft PA gel is nearly incompressible and the deformation gradient tensor can be simplified as:

$$\boldsymbol{F} = \begin{bmatrix} \beta & 0 & 0 \\ 0 & \beta & 0 \\ 2\alpha_1 \beta r_0 & 0 & \beta^{-2} \end{bmatrix} \tag{9.5}$$

The right Cauchy-Green and Green-Lagrangian finite strain tensors for large deformations are further defined as:

$$\boldsymbol{C} = \boldsymbol{F}^T \boldsymbol{F} = \begin{bmatrix} \beta^2 + 4\alpha_1^2 \beta^4 r_0^2 & 0 & 2\alpha_1 r_0 \\ 0 & \beta^2 & 0 \\ 2\alpha_1 r_0 & 0 & \beta^{-4} \end{bmatrix}; \quad \boldsymbol{E} = \frac{1}{2}(\boldsymbol{C} - \boldsymbol{I}); \tag{9.6}$$

We summarize results of radial compression ratio β and Green-Lagrangian strains in Fig. 9.5. The radial compression ratio β and the gel–water interface shape parameter α_1 were released from 0.3 to 0.8 and $O(10^{-1})$ to $O(10^{-2})$ after the violent collapse, respectively. With the assumed deformation ansatz (9.4), the $E_{\theta\theta}$ and E_{zz} components of the Green-Lagrangian strain were uniform along the curve $\widehat{ATB}$. The strain components E_{rr} and E_{rz} are location-dependent and their maximum values occurred at the bottom of the dome curve. We also found that the most severe strain occurred in the E_{zz} component, which had a maximum value of 6000% tensile-strain during the second collapse and caused the material to fail, consistent with our experimental observation that the gel–water interface was stretched to failure in the z-direction and a hole was created at the apex of the dome.

We also tracked the movement of points C and D in Fig. 9.4b, which were fitted with second order polynomials, as shown in Fig. 9.6a, b. The average propagation speed of the interface Rayleigh wave, $(|v_C| + |v_D|)/2$, attenuated linearly with time due to the viscoelasticity in the gel and took a value of $\alpha\ (G_\infty/\rho)^{1/2}$, where α was 0.5 ~ 2.3 in our experiments.

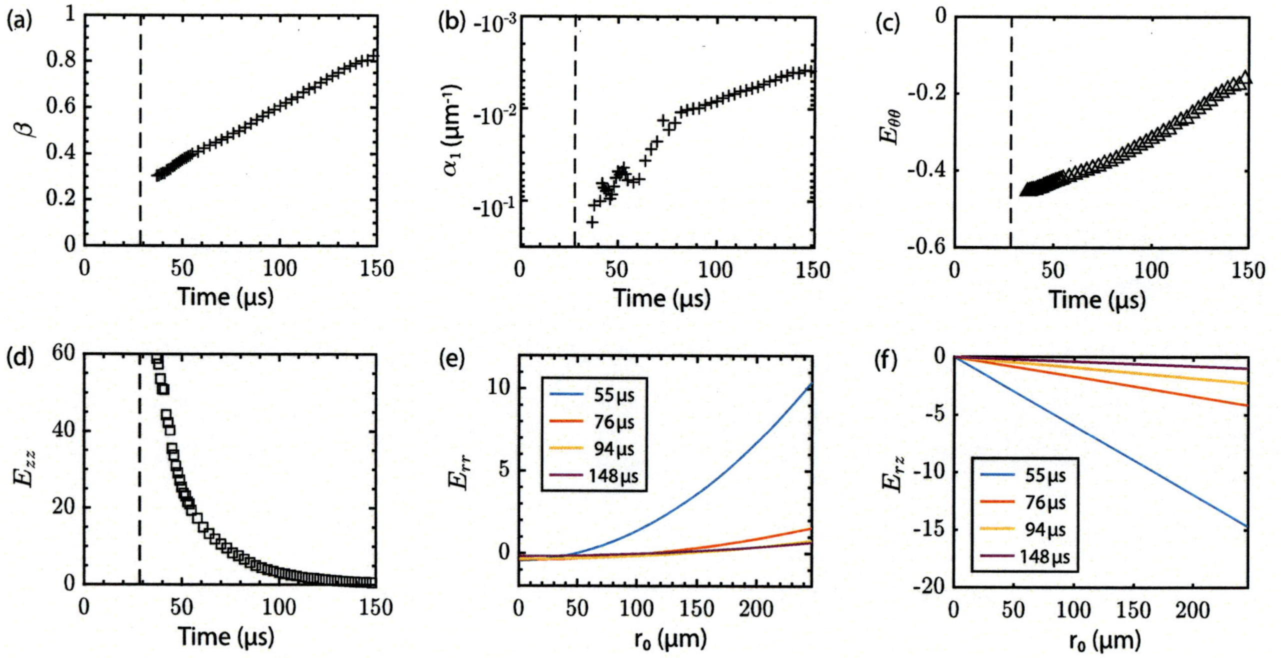

Fig. 9.5 (**a**) Radial compression ratio β. (**b**) Gel–water interface shape parameter α_1. (**c, d**) The $E_{\theta\theta}$ and E_{zz} components of the Green-Lagrangian strain. (**e, f**) The $E_{\theta\theta}$ and E_{zz} components of the Green-Lagrangian strain. (Vertical dashed lines in (**a–d**) indicate when the first violent collapse happens)

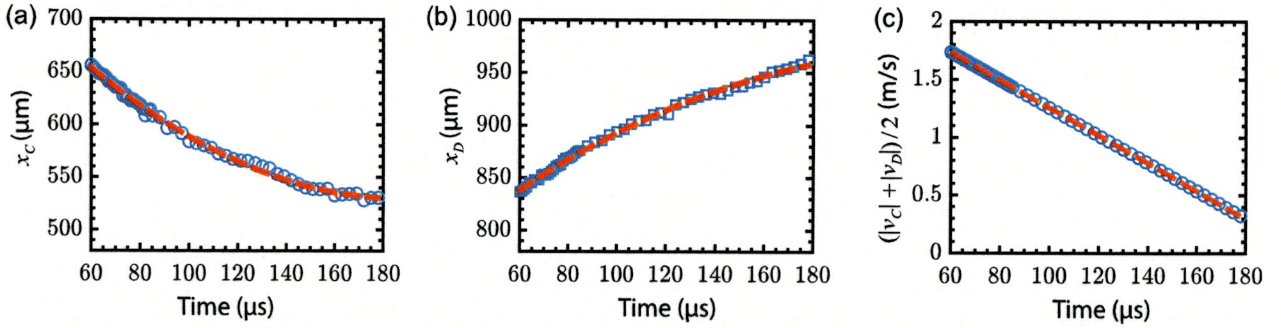

Fig. 9.6 (**a, b**) Tracked x-directional movement of points C and D in (**b**), where dashed lines are fitted second order polynomials. (**c**) The average propagation speed of the interface Rayleigh wave attenuated linearly with time due to the viscoelasticity in the gel

9.5 Conclusions

In this paper, we captured the complex interaction of an inertial cavitated bubble at a soft hydrogel–water interface. We conclude that liquid jet penetration into the boundary, jet-like ejection of the gel material, and tensile-stress-induced material failure are the major mechanisms responsible for cavitation erosion. We also find that although the bubble is located at the gel–water interface, the kinetics of its effective radius before the first collapse is almost identical to that in an isotropic gel, while the kinetics of the bubble second expansion-collapse cycle is almost identical to that in pure water. After the first violent collapse, the propagation speed of the induced interface Rayleigh wave attenuates linearly with time, due to the viscoelasticity in the gel.

Lastly, although our investigations shed new light on the intricate interaction of a bubble near a gel–fluid interface, the complex non-spherically symmetric behavior after the first collapse is however not yet fully understood. We hope that our experimental findings can motivate future studies of a fully 3D numerical simulation of the bubble gel–water interface interaction to provide a better understanding of the complicated inertial cavitation dynamics.

Acknowledgments We gratefully acknowledge the funding support from the Office of Naval Research (Dr. Timothy Bentley) under grants N00014-18-1-2625.

References

1. Brennen, C.E.: Cavitation and Bubble Dynamics. Cambridge University Press (2014)
2. Wan, M., Feng, Y., ter Haar, G.: Cavitation in Biomedicine. Springer (2015)
3. Maxwell, A.D., Wang, T.-Y., Yuan, L., Duryea, A.P., Xu, Z., Cain, C.A.: A tissue phantom for visualization and measurement of ultrasound-induced cavitation damage. Ultrasound Med. Biol. **36**, 2132–2143 (2010)
4. Mancia, L., Vlaisavljevich, E., Yousefi, N., Rodriguez, M., Ziemlewicz, T.J., Lee Jr., F.T., Henann, D., Franck, C., Xu, Z., Johnsen, E.: Modeling tissue selective cavitation damage. Phys. Med. Biol. (2019)
5. Quinto-Su, P., Dijkink, R., Prabowo, F., Gunalan, K., Preiser, P., Ohl, C.: Interaction of red blood cells with arrays of laser-induced cavitation bubbles. In: Proceedings of the 7th International Symposium on Cavitation (2009)
6. Bailey, M.R., Khokhlova, V.A., Sapozhnikov, O.A., Kargl, S.G., Crum, L.A.: Physical mechanisms of the therapeutic effect of ultrasound (a review). Acoust. Phys. **49**, 369–388 (2003)
7. Ibsen, S., Schutt, C.E., Esener, S.: Microbubble-mediated ultrasound therapy: a review of its potential in cancer treatment. Drug Des. Dev. Ther. **7**, 375 (2013)
8. Liu, Y., Zhang, A., Tian, Z., Wang, S.: Dynamical behavior of an oscillating bubble initially between two liquids. Phys. Fluids. **31**, 092111 (2019)
9. Blake, J.R., Taib, B.B., Doherty, G.: Transient cavities near boundaries. Part 1. Rigid boundary. J. Fluid Mech. **170**, 479–497 (1986)
10. Brujan, E.A., Nahen, K., Schmidt, P., Vogel, A.: Dynamics of laser-induced cavitation bubbles near an elastic boundary. J. Fluid Mech. **433**, 251–281 (2001)
11. Brujan, E.A., Nahen, K., Schmidt, P., Vogel, A.: Dynamics of laser-induced cavitation bubbles near elastic boundaries: influence of the elastic modulus. J. Fluid Mech. **433**, 283–314 (2001)
12. Yang, J., Cramer III, H.C., Franck, C.: Extracting non-linear viscoelastic material properties from violently-collapsing cavitation bubbles. Extreme Mech. Lett. **39**, 100839 (2020)
13. Estrada, J.B., Barajas, C., Henann, D.L., Johnsen, E., Franck, C.: High strain-rate soft material characterization via inertial cavitation. J. Mech. Phys. Solids. **112**, 291–317 (2018)
14. Yang, J., Franck, C.: Strain stiffening effects of soft viscoelastic materials in inertial microcavitation. In: Dynamic Behavior of Materials in Conference Proceedings of the Society for Experimental Mechanics, vol. 1, (2020)
15. Yang, J., Cramer III, H.C., Franck, C.: Dynamic rugae strain localizations and instabilities in soft viscoelastic materials during inertial microcavitation. In: Dynamic Behavior of Materials in Conference Proceedings of the Society for Experimental Mechanics, vol. 1, (2021)

Chapter 10
Effects of Hydration on the Mechanical Response of a PVA Hydrogel

Fan Cui, Jikun Wang, Alan Zehnder, and Chung-Yuen Hui

Abstract The effect of drying on the tensile behavior of a dual cross-linked poly(vinyl alcohol) (PVA) hydrogel is studied here. This gel contains about 90% water when fully hydrated. The mass–volume relationship of the gel is measured using a microbalance with a density kit. Our results show that as the gel dries the volume is linearly proportional to the mass. The impact of drying on the gel's mechanical properties is measured in uniaxial tension tests, which include loading-unloading tests at three different constant stretch rates, a complex loading history test and a stress-relaxation test. Data from specimens with different hydration levels can be described by a constitutive model of the gel. The results show that the model parameters are strongly dependent on hydration level and that as the gels dry, the gels are much stiffer than those in the fully hydrated state.

Key words Viscoelasticity · Hydration · Large strain · Dual cross-linked hydrogel

10.1 Introduction

A hydrogel is essentially a network of polymer chains swollen in water. Typical single chemical cross-linked hydrogels have poor mechanical strength limiting their applications. One method to overcome this is to introduce non-covalent, transient cross-links as a sacrificial network that can fail and then reform [1]. Hydrogels are prone to drying, leading to shrinkage and poor mechanical properties such as enhanced stiffness and reduced toughness. However, there is very little work on how dehydration affects mechanical properties. In this work, we focus on a dual-cross-linked PVA hydrogel developed by Mayumi et al. [2]. This hydrogel has both chemical and physical cross-links. The chemical cross-links form a permanent network. The physical cross-links form a temporary network which can break and reattach. For a fully hydrated gel, we and our coworkers have established a constitutive model which accurately predicts its behavior [3, 4]. Liu et al. [5] demonstrated the mechanical properties of this PVA hydrogel over a range of temperatures can also described by this constitutive model. Meacham et al. [6] studied the effect of hydration on the tensile response this PVA hydrogel. Here we report on an experimental study of the effects of drying on the constitutive response of the PVA hydrogel. We find that the same constitutive model can be used to explain our experimental results.

10.2 Experimental Methods

Experiments were performed with a dual-cross-linked PVA hydrogel which contained approximately 12% PVA and the balance an ionic solution. The gel was synthesized in three steps: making a 16% PVA solution, adding chemical cross-links, then adding physical cross-links. First, PVA powder was added to distilled water at 5 °C. The mixture was stirred and heated to 95 °C to dissolve the PVA in the water. Then the PVA solution and glutaraldehyde cross-linker were mixed with hydrochloric acid. The solution was injected in mold and held for 24 h to form the chemical cross-links. Finally, the chemically cross-linked gel was washed to neutralize the pH by soaking in distilled water for 24 h. It was then soaked in

F. Cui · J. Wang · A. Zehnder (✉) · C.-Y. Hui
Mechanical and Aerospace Engineering, Cornell University, Ithaca, NY, USA
e-mail: fc343@cornell.edu; jw2586@cornell.edu; atz2@cornell.edu; ch45@cornell.edu

A. Amirkhizi et al. (eds.), *Challenges in Mechanics of Time Dependent Materials, Mechanics of Biological Systems and Materials & Micro-and Nanomechanics, Volume 2*, Conference Proceedings of the Society for Experimental Mechanics Series,
https://doi.org/10.1007/978-3-030-86737-9_10

an ionic solution made of water, sodium chloride and borax to form the ionic (physical) cross-links. After 3–4 days, the physical bonds have completely formed, and the gel is ready to be used in experiments.

The sample was then allowed to dry in lab air. Hydration levels were characterized by the amount of mass loss during drying. We also measured mass-volume relationship at different hydration level. The mass and volume of a fully hydrated sample were measured initially. We then measured its mass every 5–10 min until it reached 95%, 90%, 85%, 80%, 75%, 70%, 65%, and 60% of the initial mass. For each of these hydration levels we also measured its volume. The mass was measured by a microbalance with an accuracy of 0.01 mg. The volume was measured by weighing the sample in air and in mineral oil using a density kit on the microbalance. The difference between these two values is the buoyancy which allows us to calculate the volume of the sample via Archimedes principle and the oil density.

The mechanical tests were performed with a custom-built tester using a stepper motor driven translation stage, a 1 N load cell, an oil container, LVDT and DVM-based data acquisition system [7]. Samples were cut into strips 2 mm thick, 10 mm wide initially and clamped between a pair of sandpaper lined aluminum grips separated by a gauge length of approximately 30 mm. All tests were performed in mineral oil to prevent the sample from further drying.

At each hydration level (100, 90, 80, 70, and 60%), six tests were performed. Tests were performed sequentially using the same sample for consistency. Between each test, the sample was allowed relaxed for about 12 min to fully recover to its initial state. We carried out three different types of tests. First, we carried out cyclic tests where the sample was loaded to a stretch of $\lambda = 1.3$ at stretch rates of $\dot{\lambda} = 0.003/\mathrm{s}, 0.010/\mathrm{s}, 0.03/\mathrm{s}$ and then unloaded at the same rate. Second, we carried out a complex loading history in which the sample was loaded to a stretch of $\lambda = 1.15$ at a rate of $\dot{\lambda} = 0.003/\mathrm{s}$, held for 1 min and then loaded to $\lambda = 1.3$ at $\dot{\lambda} = 0.030/\mathrm{s}$, held for another 1 min and then unloaded at a rate of $\dot{\lambda} = 0.010/\mathrm{s}$. Then the sample was loaded to $\lambda = 1.3$ at $\dot{\lambda} = 0.100/\mathrm{s}$ and unloaded at $\dot{\lambda} = 0.001/\mathrm{s}$. This is followed by a tensile-relaxation test at a maximum stretch of $\lambda = 1.3$ with an initial stretch rate of $\dot{\lambda} = 0.500/\mathrm{s}$.

Results are presented in terms of the nominal stress, i.e., applied force divided by the initial cross-sectional area of the sample. For the fully hydrated gel, the cross-sectional area was 2 mm × 10 mm. For the drying gels, the gel shrinks uniformly in all three dimensions, thus the cross-sectional area is calculated using 2 mm × 10 mm × $(V\%)^{2/3}$, where $V\%$ is the volume percentage (volume after drying divided by initial volume at full hydration). This relationship is found from the mass–volume relationship described below.

10.3 Experimental Results

The results of mass–volume experiments are shown in Fig. 10.1. The volume percentage decreases linearly with the mass percentage (mass after drying divided by initial mass). The data in Fig. 10.1 can be fit using $V\% = 1.03M\% - 3.04\%$. This equation allows us to calculate the cross-section area of the gel at different hydration levels as described above.

Figures 10.2, 10.3, and 10.4 show the stress–strain curves for PVA hydrogels at different hydration levels for the first two types of tensile tests (cyclic and complex loading). Figure 10.5 shows the stress–time curves for PVA hydrogels at different hydration levels in tensile-relaxation tests.

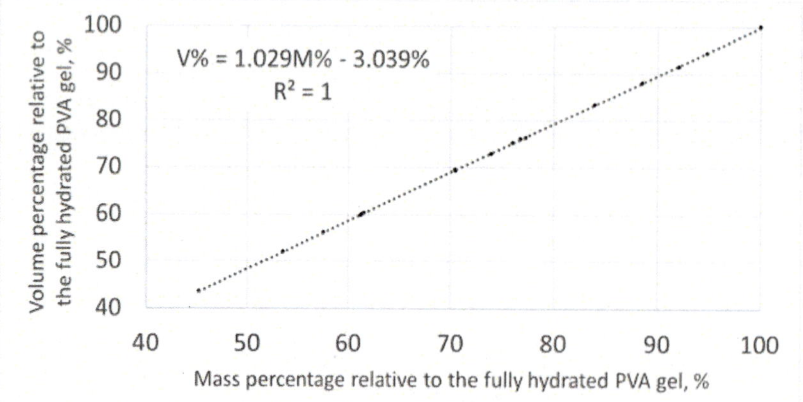

Fig. 10.1 Mass–volume relationship of PVA hydrogels as they dry

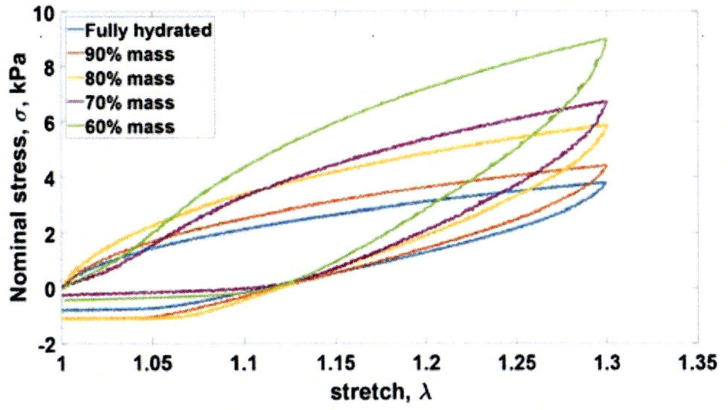

Fig. 10.2 Nominal stress vs. stretch at five hydration states. Load-unload at stretch rate of $\dot{\lambda} = 0.003/\mathrm{s}$

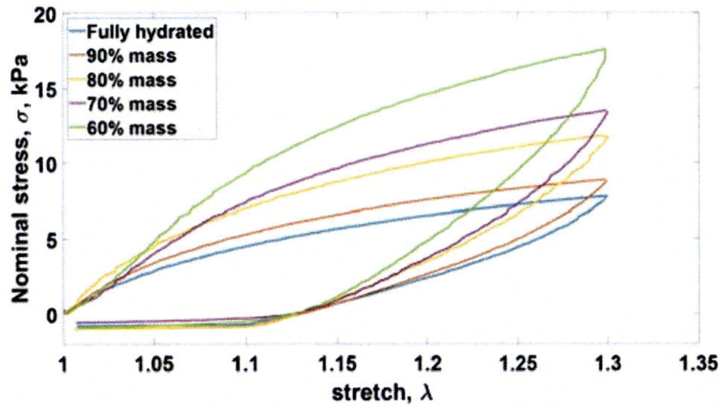

Fig. 10.3 Nominal stress vs. stretch at five hydration states. Load-unload at stretch rate of $\dot{\lambda} = 0.030/\mathrm{s}$

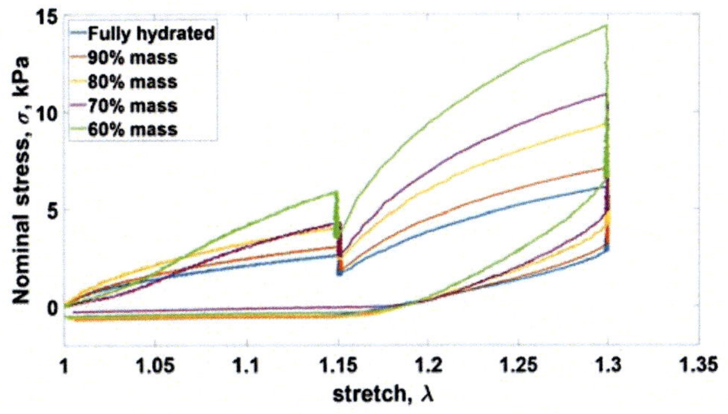

Fig. 10.4 Nominal stress vs. stretch at five hydration states. A complex loading history

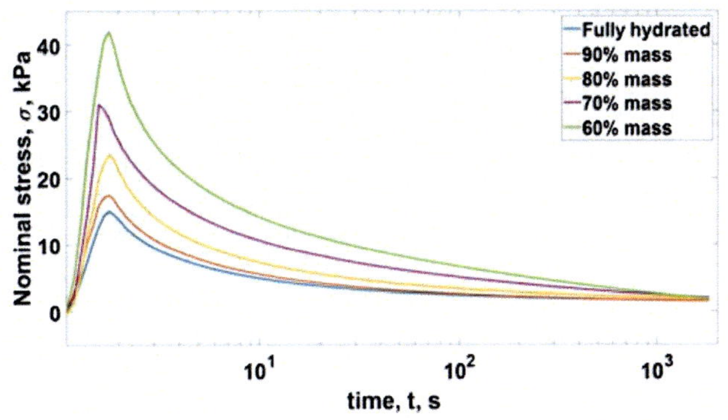

Fig. 10.5 Nominal stress vs. time in log scale at five hydration states. Loading rate is 0.5/s. Maximum stretch is 1.3

These figures show that for all stretch rates the stress is significantly higher rate as the gel dries. Exceptions are gels at 70% and 60% mass. Figure 10.2 shows that the stress increases slowly at the initial part of loading for these hydration levels. One possible reason is that gel samples with mass percentage less than 70% have not had sufficient time to recover after the previous experiment. Even with this initial dip, the maximum nominal stress of the drier gels is higher and across all tests we see that the gel stiffens as it dehydrates.

10.4 Data Fitting and Discussion

The constitutive model of PVA dual cross-linked hydrogel was developed by Long et al. [3] and Guo et al. [4]. For uniaxial loading the nominal stress σ is related to stretch λ by

$$\sigma = \mu(\rho + n(t))\left[\lambda(t) - \frac{1}{\lambda^2(t)}\right] + \mu\overline{\gamma}_\infty \times \int_0^t \phi_B\left(\frac{t-\tau}{t_B}\right)\left[\frac{\lambda(t)}{\lambda^2(\tau)} - \frac{\lambda(\tau)}{\lambda^2(t)}\right]d\tau, \tag{10.1}$$

where

μ is the small strain shear modulus of neo-Hookean model,
ρ is the molar fraction of the chemical cross-links,
$\overline{\gamma}_\infty$ is the steady state reattachment rate of the temporary chains, i.e., molar fraction of the temporary chains reattached per unit time, and

$$\phi_B\left(\frac{t-\tau}{t_B}\right) = \left(1 + (\alpha_B - 1)\frac{t}{t_B}\right)^{\frac{1}{1-\alpha_B}}, \tag{10.2}$$

which denotes the probability of breaking of physical cross-links, where t_B is the characteristic time for breaking; $2 > \alpha_B > 1$ is a material constant that specifies the rate of decay of ϕ_B.

$n(t)$ is the fraction of physical bonds present at $t = 0$ and still attached at t, which is

$$n(t) = \overline{\gamma}_\infty \int_{-\infty}^0 \phi_B\left(\frac{t-\tau}{t_B}\right)d\tau = \overline{\gamma}_\infty \frac{t_B}{2-\alpha_B}\left(1 + (\alpha_B - 1)\frac{t}{t_B}\right)^{\frac{2-\alpha_B}{1-\alpha_B}}, \tag{10.3}$$

The constitutive model has four independent parameters, $\mu\rho$, $\mu\overline{\gamma}_\infty$, α_B and t_B.

We use a machine learning algorithm to determine material parameters. Figure 10.6 shows that these parameters provide very accurate predictions which fit the experimental results for a fully hydrated PVA dual-cross-linked hydrogel. For other hydration levels, the agreement between the model prediction and experiment are just as good. The best fit model parameters for the PVA hydrogel at different hydration levels are given in Table 10.1.

According to the constitutive model, $\mu\rho$ describes the long-time response of the chemical network in a relaxation test. When the PVA gel dries from 100% mass to 60% mass, $\mu\rho$ increases by about 47%. This means the chemical network carries more load when the gel dries. Table 10.1 shows that α_B changes little when the gel dehydrates while t_B decreases and $\mu\overline{\gamma}_\infty$ increases. This suggests physical bonds break and reattach faster when the gel loses water. The last column in the table, $\frac{\mu\overline{\gamma}_\infty t_B}{2-\alpha_B}$, can be seen as a parameter that describes the short-time response of physical bonds at small strains. It grows three times higher when the PVA gel dries from 100% mass to 60% mass. Thus the short-time modulus of the gel, which involves both the physical bonds and chemical network, increases strongly with drying. The drier PVA gels are stiffer and water content has a much more significant influence on the physical cross-links than chemical cross-links as evidenced by the far greater increase in $\frac{\mu\overline{\gamma}_\infty t_B}{2-\alpha_B}$ relative to $\mu\rho$.

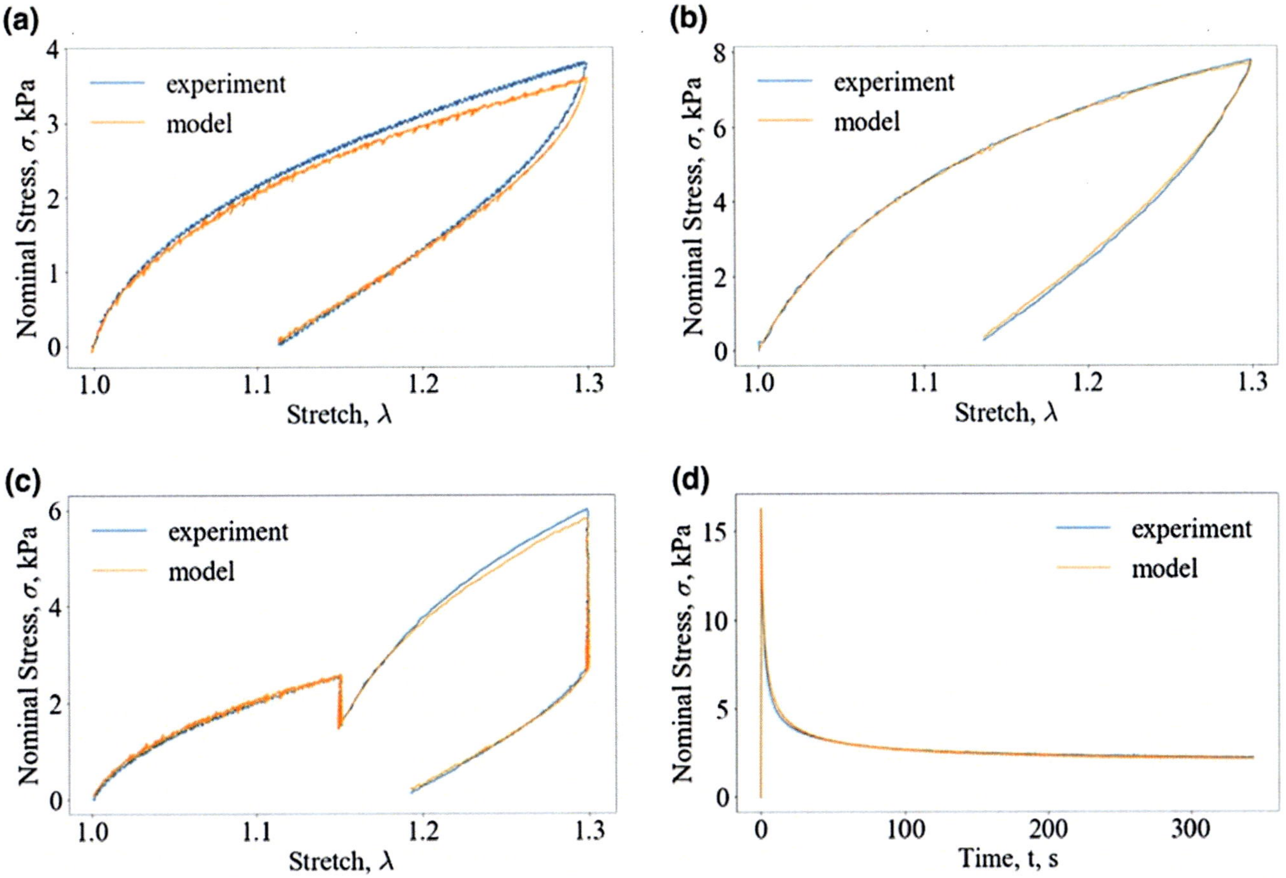

Fig. 10.6 Fitting results for fully hydrated (100% mass) PVA hydrogel. Nominal Stress vs. Stretch for (**a**) Loading-unloading to stretch $\lambda = 1.3$ at stretch rate 0.003/s, (**b**) Loading-unloading to stretch $\lambda = 1.3$ at stretch rate 0.03/s, (**c**) A complex loading history which was mentioned above, and (**d**) Nominal Stress vs. Time for loading at stretch rate 0.5/s to a stretch of $\lambda = 1.3$ and then holding as the stress relaxes

Table 10.1 Data fitting results for PVA hydrogels at different hydration levels

Hydration	μ_P(kPa)	α_B	$t_B(s)$	$\mu\bar{\gamma}_\infty$(kPa/s)	$\frac{\mu\bar{\gamma}_\infty t_B}{2-\alpha_B}$ (kPa)
100% mass	2.347	1.634	0.4618	19.91	25.15
90% mass	2.567	1.633	0.4335	26.72	31.53
80% mass	2.640	1.664	0.4250	31.62	40.05
70% mass	2.994	1.674	0.3383	48.76	50.62
60% mass	3.439	1.687	0.2338	101.2	75.55

Acknowledgments This material is based upon work supported by the National Science Foundation, under Grant No. CMMI-1903308.

References

1. Gong, J., Katsuyama, Y., Kurokawa, T., Osada, Y.: Double-network hydrogels with extremely high mechanical strength. Adv. Mater. **15**(14), 1155–1158 (2003)
2. Mayumi, K., Marcellan, A., Ducouret, G., Creton, C., Narita, T.: Stress strain relationship of highly stretchable dual cross-link gels: separability of strain and time effect. ACS Macro Lett. **2**(12), 1065–1068 (2013)
3. Long, R., Mayumi, K., Creton, C., Narita, T., Hui, C.-Y.: Time dependent behavior of a dual cross-link self-healing gel: theory and experiments. Macromolecules. **47**, 7243–7250 (2014)

4. Guo, J., Long, R., Mayumi, K., Hui, C.-Y.: Mechanics of a dual cross-link gel with dynamic bonds: steady state kinetics and large deformation effects. Macromolecules. **49**, 3497–3507 (2016)
5. Liu, M., Guo, J., Hui, C.-Y., Creton, C., Narita, T., Zehnder, A.: Time-temperature equivalence in a PVA dual cross-link self-healing hydrogel. J. Rheol. **62**(4), 991–1000 (2018)
6. Meacham, R., Liu, M., Guo, J., Zehnder, A., Hui, C.-Y.: Effect of hydration on tensile response of a dual cross-linked PVA hydrogel. Exp. Mech. **60**(8) (2020)
7. Liu, M., Guo, J., Hui, C.-Y., Zehnder, A.: Crack tip stress based kinetic fracture model of a PVA dual-crosslink hydrogel. Extreme Mech. Lett. **29** (2019)

Chapter 11

Gaussian Process to Identify Hydrogel Constitutive Model

Jikun Wang, Tianjiao Li, Chung-Yuen Hui, Jingjie Yeo, and Alan Zehnder

Abstract Unlike traditional structural materials, soft solids can often sustain very large deformation before failure, and many exhibit nonlinear viscoelastic behavior. Modeling nonlinear viscoelasticity is a challenging problem for a number of reasons. In particular, a large number of material parameters are needed to capture material response and validation of models can be hindered by limited amounts of experimental data available. We have developed a Gaussian Process (GP) approach to determine the material parameters of a constitutive model describing the mechanical behavior of a soft, viscoelastic PVA hydrogel. A large number of stress histories generated by the constitutive model constitute the training sets. The low-rank representations of stress histories by Singular Value Decomposition (SVD) are taken to be random variables which can be modeled via Gaussian Processes with respect to the material parameters of the constitutive model. We obtain optimal material parameters by minimizing an objective function over the input set. We find that there are many good sets of parameters. Further the process reveals relationships between the model parameters. Results so far show that GP has great potential in fitting constitutive models.

Key words Viscoelastic · Tension test · Finite deformation · Large strain · Stress relaxation · Machine learning

11.1 Introduction

Many materials, for example, soft solids such as rubber and gels, are viscoelastic or viscoplastic solids. These materials, besides being rate-sensitive, can sustain very large deformation before failure. To describe their mechanical properties, researchers have designed many complex constitutive models. However, most constitutive models contain many material parameters that cannot be directly determined by experimental data. In addition, validating the theoretical model is hindered by the limited amount of experimental data available. The process of fitting experimental data to theory can be extremely tedious and time-consuming. When the number of material parameters is large, poor fitting of data can occur even if the model captures the correct physics. To our best knowledge, it is still a great challenge to rapidly determine the parameters for complex viscoelastic constitutive models. Here we propose a method to find the parameters for constitutive models, which combines singular value decomposition (SVD) and machine learning tools, specifically, Gaussian process. Although our formulation is general, we demonstrate its usage and validate our algorithm by applying it to study the mechanical behavior of a nonlinear viscoelastic PVA hydrogel.

11.2 PVA Constitutive Model

In our previous works [1], we have developed a 3D constitutive model which combines the finite strain elasticity of elastomers with the kinetics of bond breaking and reattachment. We have also demonstrated that our model accurately predicts results from uniaxial tension and torsion tests with complex loading histories. The constitutive model for the PVA gel is completely

J. Wang · T. Li · C.-Y. Hui · J. Yeo · A. Zehnder (✉)
Mechanical and Aerospace Engineering, Cornell University, Ithaca, NY, USA
e-mail: jw2586@cornell.edu; tl787@cornell.edu; ch45@cornell.edu; jingjieyeo@cornell.edu; atz2@cornell.edu

A. Amirkhizi et al. (eds.), *Challenges in Mechanics of Time Dependent Materials, Mechanics of Biological Systems and Materials & Micro-and Nanomechanics, Volume 2*, Conference Proceedings of the Society for Experimental Mechanics Series,
https://doi.org/10.1007/978-3-030-86737-9_11

determined by four material parameters $\mu\rho$, α_B, t_B and $\mu\overline{\gamma}_\infty$. Hence a parameter set is specified by a four-component vector $\vec{x} = (\mu\rho, \alpha_B, t_B, \mu\overline{\gamma}_\infty)$. According to our constitutive model, in a uniaxial tension test where the stretch ratio $\lambda(t)$ is prescribed, the nominal stress $\sigma(t)$ corresponding to the parameter set $\vec{x} \equiv (\mu\rho, t_B, t_B, \mu\overline{\gamma}_\infty) \in \Omega \subset \mathbb{R}^4$ is

$$\sigma(t) = \left[\mu\rho + \mu\overline{\gamma}_\infty \frac{t_B}{2-\alpha_B}\left(1 + (\alpha_B - 1)\frac{t}{t_B}\right)^{\frac{2-\alpha_B}{1-\alpha_B}}\right]\left[\lambda(t) - \frac{1}{\lambda^2(t)}\right] + \mu\overline{\gamma}_\infty \int_0^t \phi_B\left(\frac{t-\tau}{t_B}\right)\left[\frac{\lambda(t)}{\lambda^2(\tau)} - \frac{\lambda(\tau)}{\lambda^2(t)}\right] d\tau \tag{11.1a}$$

where

$$\phi_B\left(\frac{t}{t_B}\right) = \left(1 + (\alpha_B - 1)\frac{t}{t_B}\right)^{\frac{1}{1-\alpha_B}} \tag{11.1b}$$

The units of parameters are $(\mu\rho, \alpha_B, t_B, \mu\overline{\gamma}_\infty)$(kPa, 1, s, kPa).

11.3 Gaussian Process Machine Learning

Singular value decomposition (SVD) and Gaussian process have been used by different groups of researchers to predict material behavior or to find crucial parameters for controlling system behavior [2, 3]. Here we use these tools to help us predict the output of our constitutive model, i.e., stress history. For a fixed strain history, we can use the constitutive model to calculate the stress history for different parameters. If we put all those stress histories together, we can get a stress matrix and apply singular value decomposition on it to get the basis and principal components of each stress history. Then we use the principal components of all those stress histories to train a Gaussian process. After training, under the same strain history, this Gaussian process can make predictions about the principal components of the stress history of any parameters it has not seen. As a result, we can use GP to evaluate the stress history over a large number of parameters then find the parameters that best fit the experimental data. The flowchart of Gaussian process machine learning is shown in Fig. 11.1.

11.4 Experimental Methods

The dual-cross-link poly(vinyl alcohol) (PVA) hydrogels were prepared by incorporating ions in a chemically crosslinked PVA gel. This hydrogel system was first introduced by Mayumi et al. [4]. First a 16 wt% PVA solution was made by dissolving PVA powder in distilled water. Then the PVA solution was chemically cross-linked by glutaraldehyde in an acidic environment. Then the chemical gel was washed several times to neutralize the pH. Finally, the gel was soaked in Borax/NaCl solution to form the physical bonds. The gel was soaked in the solution for at least 3 days to reach equilibrium prior to testing.

We performed uniaxial tension tests using a single PVA hydrogel specimen with four different strain histories. Some of the physical bonds will be broken after a test. Fortunately, the dual-crosslink PVA hydrogel has self-healing property; the bonds can reform and the gel's mechanical properties can totally recover to their original state after 30 min at room temperature.

The tests were performed using a custom-built tensile tester with the samples immersed in a mineral oil bath to prevent the specimens from drying. We conducted EXP 1, EXP 2, EXP 3, and EXP 4 sequentially, as shown in Fig. 11.2. After each test, we kept the specimen in the oil bath 30 min to let it recover to the as-prepared state.

11.5 Results and Discussion

The best fitting result given by GP is illustrated in Fig. 11.3. The agreement between experiments and theory is excellent. These results further demonstrate that our PVA constitutive model correctly captures the mechanical behavior of PVA gels. In addition, it shows that our machine learning algorithm is a powerful tool for determining material parameters in constitutive models.

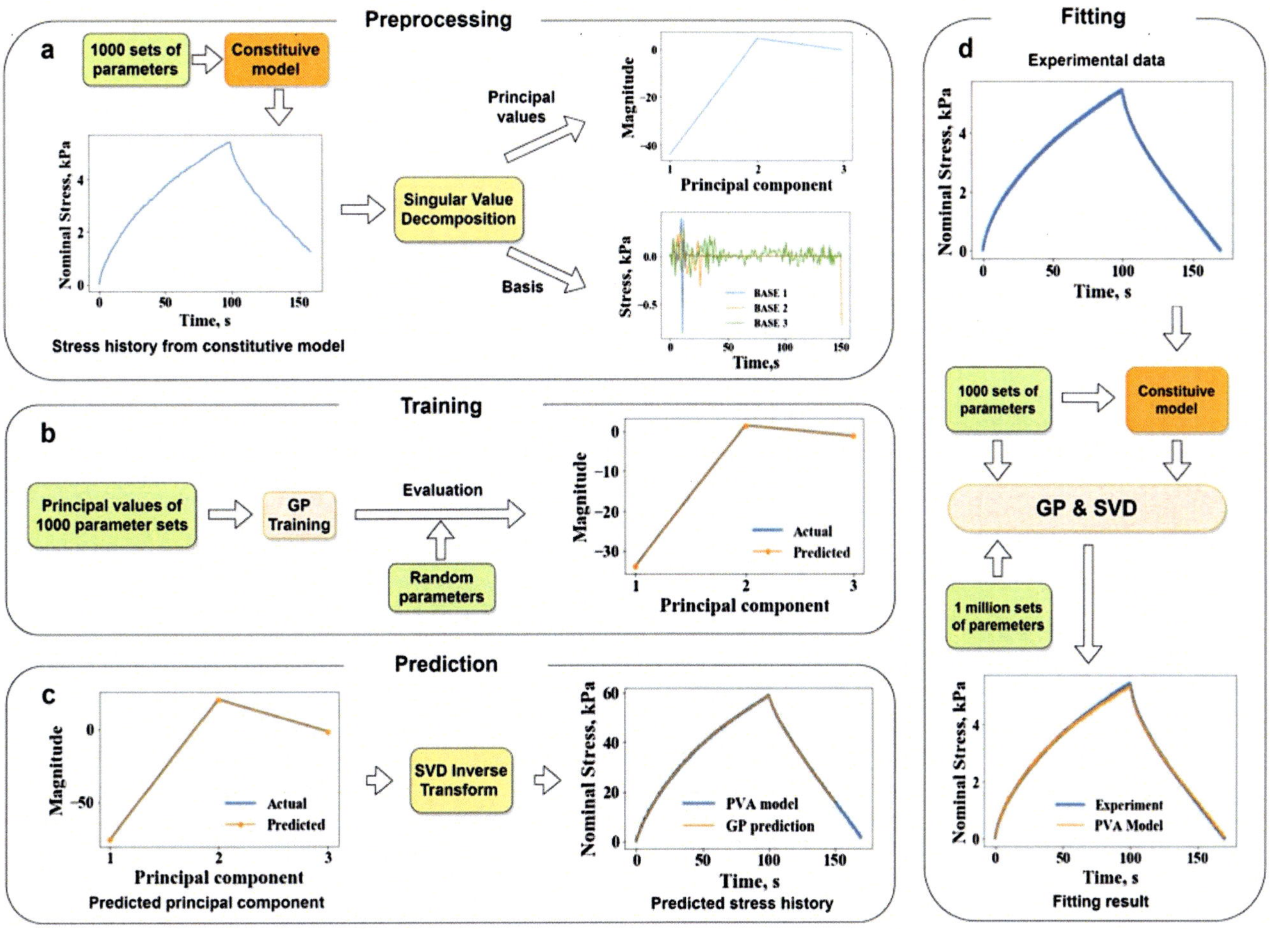

Fig. 11.1 Workflow of Gaussian Process to identify constitutive model. **(a)** 1000 stress history curves calculated directly from the constitutive model are decomposed into basis and projections on this basis (i.e., principal components) using SVD; **(b)** the principal components of the 1000 parameter sets are taken as training set. After training, the GP can make predictions about the principal components of any parameter set not in the training set; **(c)** with the predicted principal components, the stress history of any parameter set can be predicted by applying SVD inverse transform to the predicted principal component without calculating the constitutive model; **(d)** the best parameters for the experimental data can be obtained by comparing the experimental stress history with the stress histories calculated for a large number of parameter sets

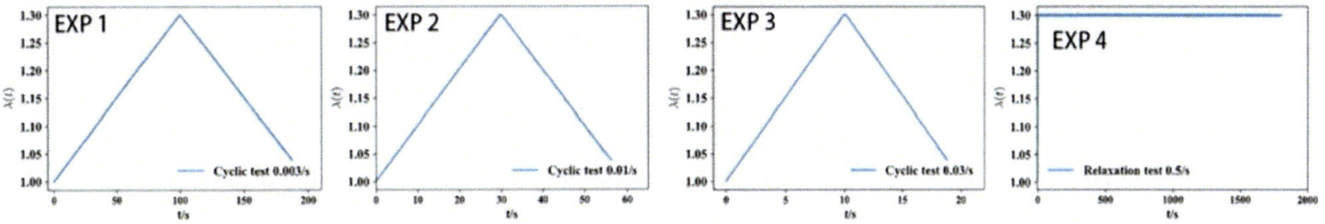

Fig. 11.2 Four loading histories for PVA gels

Furthermore, using our method, we discovered that there are many sets of material parameters that fit the experiments well. The projections of those parameters on 2D subspaces are drawn in Fig. 11.4, and these plots show that there are strong relationships between $\mu\rho$ and α_B, as well as t_B and $\mu\overline{\gamma}_\infty$. To understand the relationships, we use a result from our previous work [1], which demonstrates that when strains are small, that is, when $\lambda(t) \approx 1 + \epsilon(t)$, the constitutive model becomes

$$\sigma(t) = \int_{-\infty}^{t} Y\left(\frac{t-\tau}{t_B}\right)\frac{d\epsilon(\tau)}{d\tau}d\tau \tag{11.2a}$$

where

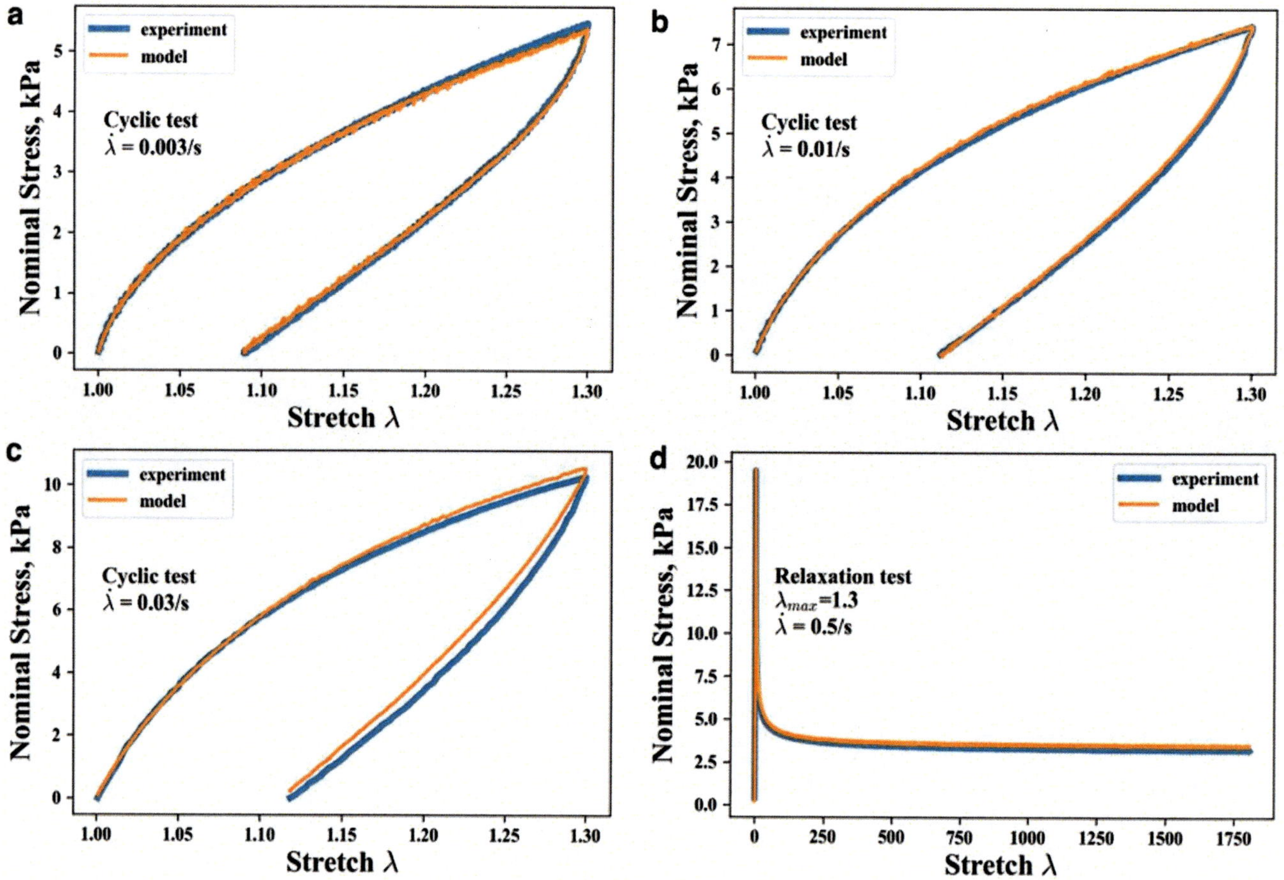

Fig. 11.3 Comparison of model prediction to experiments using the parameters given by GP. **(a)** Experiment 1, **(b)** Experiment 2, **(c)** Experiment 3, and **(d)** Experiment 4

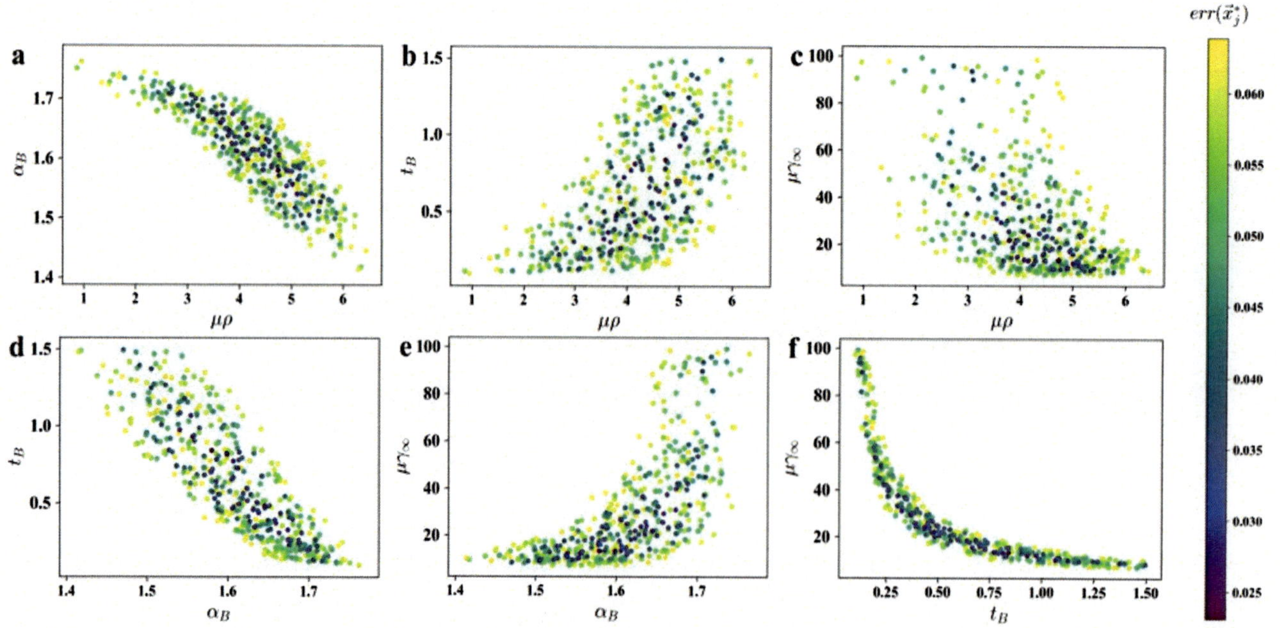

Fig. 11.4 Distribution of 500 parameter sets which fit the experiments very well. **(a)** α_B vs. $\mu\rho$, **(b)** t_B vs. $\mu\rho$ **(c)** $\mu\gamma_\infty$ vs. $\mu\rho$, **(d)** t_B vs. α_B, **(e)** $\mu\gamma_\infty$ vs. α_B, **(f)** $\mu\gamma_\infty$ vs. t_B

$$Y(t) \equiv \frac{3\mu\overline{\gamma}_{\infty}t_B}{2-\ \alpha_B}\left(1+(\ \alpha_B-1)\frac{t}{t_B}\right)^{\frac{2-\alpha_B}{1-\alpha_B}}+3\mu\rho \tag{11.2b}$$

From this equation, it is clear that $\sigma(t)$ depends linearly on $\mu\overline{\gamma}_{\infty}$. Then $\mu\overline{\gamma}_{\infty}t_B$ must be a material constant C for a specific experiment, therefore $\mu\overline{\gamma}_{\infty} = C/t_B$. And it is clear that $\sigma(t)$ increases with $\mu\rho$ and α_B. Therefore, to produce the same stress history, a smaller α_B must be chosen if a larger value of $\mu\rho$ is already chosen. Figure 11.4 shows the machine learning predictions are consistent with those relationships, which further demonstrates the power of our method.

Theoretically, all the results above can be obtained by calculating the constitutive model for one million parameter sets directly, however, a prediction that would take 1.6 min using SVD and Gaussian process metamodel for one million parameter vectors would take up to 322 h to evaluate using the constitutive model. For more complicated constitutive models and longer strain histories, it is impractical to calculate the stress history for millions of parameter sets. This is the main advantage of our method over the traditional methods of data fitting.

Acknowledgments This material is based upon work supported by the National Science Foundation, under Grant No. CMMI-1903308 and Grant No. CMMI-2038057.

References

1. Guo, J., Long, R., Mayumi, K., Hui, C.Y.: Mechanics of a dual cross-link gel with dynamic bonds: steady state kinetics and large deformation effects. Macromolecules. **49**, 3497–3507 (2016)
2. Guo, M., Hesthaven, J.S.: Data-driven reduced order modeling for time-dependent problems. Comput. Methods Appl. Mech. Eng. **345**, 75–99 (2019)
3. Yang, C., Kim, Y., Ryu, S., Gu, G.X.: Prediction of composite microstructure stress-strain curves using convolutional neural networks. Mater. Des. **189**, 108509 (2020)
4. Mayumi, K., Marcellan, A., Ducouret, G., Creton, C., Narita, T.: Stress strain relationship of highly stretchable dual cross-link gels: separability of strain and time effect. ACS Macro Lett. **2**, 1065–1069 (2013)

Chapter 12
Effect of Host Surface Factors on Biocompatible Adhesion Index

James D. Boyd and Martha E. Grady

Abstract Biofilm formation is a significant problem in America, accounting for 17 million infections, and causing 550,000 deaths annually. An understanding of factors that contribute to strong biofilm surface adhesion at implant interfaces can guide the development of surfaces that prevent deleterious biofilms and promote osseointegration. The aim of this research is to develop a metric that quantifies the adhesion strength differential between a bacterial biofilm and an osteoblast-like cell monolayer to a medical implant-simulant surface. This metric will be used to quantify the biocompatible effect of implant surfaces on bacterial and cell adhesion. The laser spallation technique employs a high-amplitude short-duration stress wave to initiate spallation of biological films. Attenuation of laser energy results in failure statistics across increasing fluence values, which are calibrated via interferometry to obtain interface stress values. Several metrology challenges were overcome including how membrane tension may influence laser spallation testing and how to determine stress wave characteristics when surface roughness precludes in situ displacement measurements via interferometry. Experiments relating loading region within biofilm to centroid of biofilm revealed that location played no role in failure rate. A reflective panel was implemented to measure stress wave characteristics on smooth and rough titanium, which showed no difference in peak compressive wave amplitude. After overcoming these metrology challenges, the adhesion strength of *Streptococcus mutans* biofilms and MG 63 monolayers on smooth and rough titanium substrates is measured. An Adhesion Index is developed by obtaining the ratio of cell adhesion to biofilm adhesion. This nondimensionalized parameter represents the effect of surface modifications on increases or decreases in biocompatibility. An increase in Adhesion Index value is calculated for roughened titanium compared to smooth titanium. The increase in Adhesion Index values indicates that the increase in surface roughness has a more positive biological response from MG 63 than does *S. mutans*. In this work further experiments quantifying impact of various surface coating including blood plasma, and adhesion proteins found within the extracellular matrix to expand the Adhesion Index.

Key words Biofilms · Laser spallation · Adhesion · *Streptococcus mutans* · MG 63 · Blood plasma · Surface treatment

The focus of this study is to examine the effect of improved in vivo conditions experienced during the medical implantation process, specifically, in vivo conditions experienced by dental implant devices. During implantation the first thing to reach the surface is blood which permeates the active wound site during coagulation and hemostasis [1]. Proteins found within the blood will coat the newly implanted dental device before future osteoblastic cells adhere [2]. In order to model the competition more accurately between cellular and bacterial adhesion and the implantation an inclusion of the blood plasma proteins associated with wound healing are critical. During this competition the early colonizing bacteria are the foundational microbes which promote the adhesion and growth of more pathogenic bacteria onto the newly implanted surfaces [3]. Preventing the initial adhesion of these early colonizers can drastically diminish infection rates. These early colonizing bacteria are primarily dominated by oral *Streptococcus*, making up nearly 80% of the early biofilm constituents [4]. As such to model these typical early colonizing bacteria, *Streptococcus mutans* was selected in this study. *S. mutans* is associated with dental caries and promotes the adhesion of other bacteria [3]. Dental implants are susceptible to adherent bacteria simultaneous to the natural wound healing process that begins when the titanium implant is placed in bone, i.e., as osteoblasts contribute to osseointegration [5]. The body's natural cellular response is also modeled to accurately assess the competition between invasive bacterial adhesion and host osteoblast cells, bone cells. MG 63 osteosarcoma cells were selected to model the

J. D. Boyd (✉) · M. E. Grady
Department of Mechanical Engineering, College of Engineering, University of Kentucky, Lexington, KY, USA
e-mail: jdbo227@g.uky.edu; m.grady@uky.edu

A. Amirkhizi et al. (eds.), *Challenges in Mechanics of Time Dependent Materials, Mechanics of Biological Systems and Materials & Micro-and Nanomechanics, Volume 2*, Conference Proceedings of the Society for Experimental Mechanics Series,
https://doi.org/10.1007/978-3-030-86737-9_12

behavior of immature osteoblast cells, for their excellent immature osteoblastic traits [6]. The scaffold for these adhesion studies was chosen to be commercially pure titanium, as titanium is the most commonly used dental implant material [7]. Dental implants contain both smooth and roughened titanium surfaces, roughened titanium is shown to increase cellular adhesion, thus both roughened and smooth titanium substrates were examined in order to mimic the varying surface roughnesses found on dental implants [8]. The dental implant mimicking substrate is first coated with a physiosorbed layer of blood plasma before inoculated with either the *S. mutans* or MG 63 cells. *S. mutans* bacteria are cultured in Todd Hewitt Yeast with sucrose, while MG 63 is cultured in Eagle's Minimum Essential Medium, until confluent biofilms and monolayers are produced, respectively. The dental implant mimicking assemblies, cultured with appropriate films, undergo laser spallation experiments to determine the sample-biofilm interfacial adhesion strength. The effect of surface morphology on adhesion has previously been examined using the laser spallation technique [9], but the inclusion of blood plasma coating provides an improved biometric for biocompatibility of these smooth and roughened surfaces. The adhesion strength values for both films on the smooth and rough surfaces are directly comparable values, additionally the adhesion strengths with and without plasma coating can be compared. The uncoated smooth surface titanium is used as a baseline and the effect of surface roughness with blood plasma coating is examined. Optimal implantology designs would look for surfaces which promote host cellular adhesion while deterring invasive bacterial adhesion. A previously developed metric known as the Adhesion Index examines the ratio of bacterial and cellular adhesion to determine the biocompatibility of various surface morphologies [9]. The impact blood plasma protein coatings have on existing Adhesion Index metrics are examined in this study in order to determine the role they play on both bacterial and cellular adhesion. The goal being to improve this existing metric to allow for better predictive in vitro metrics in implantology, so that in vivo results can be predicted and improved.

The most commonly used adhesion technique used in bacterial studies are counting methods [10]. These counting methods do not give any information on mechanical adhesion of biological films and direct comparison of these results is difficult because of the size scale associated bacteria vs cells. Jet impingement and shear flow studies also have been performed to quantify the adhesion of bacterial and cellular films, however these tests rely on contact methods, fluid flow, to induce stresses and can often lead to deformation of the films before accurate measurements can be obtained [11]. The low cohesiveness of these films can cause them to fail cohesively during testing before adhesion can be measured accurately [12]. Laser spallation is a thin film adhesion test that previously has been used to quantify the adhesion of metallic films onto substrates [13, 14]. More recently the technique has been employed to quantitatively measure the adhesion strength of several biological films [9, 15–17]. The laser spallation setup used is illustrated in Fig. 12.1a. Laser spallation operates by delivering a photoacoustic shockwave to the biofilm which, with sufficient energy, will spall the films from the interface surface. Calibration experiments are performed following spallation experiments in order to obtain stress wave generation at the interface for interface strength values. Interferometry is applied to measure the free surface displacement during loading to quantify and observe the substrate stress magnitude and profile. Wave transmission and reflection equations are applied to determine the final adhesion strength for films. This protocol is highlighted in greater detail in Boyd et al. [9]. The rapid non-contact stress applied by the laser spallation method is superior to other film adhesion testing methods because of the preclusion of biological film deformation over time from the low cohesiveness of the films. Additionally multiple loaded regions can be applied over a single biofilm to increase results from a single biofilm [18].

Dental implants include smooth and rough titanium surfaces, the threads of dental implants are roughened to increase osseointegration for years, while the gingival portion of the implant is often smooth to prevent irritation and oral bacteria adhesion. To mimic these surface characteristics glass slides are purchased with e-beam evaporated 100 nm coatings of titanium. The opposite side of the slide is coated in 300 nm of aluminum which acts as the energy absorbing layer for the laser used. A separate set of slides are first sandblasted with large grit to obtain a surface roughness between 1 and 1.5 μm. The glass slides are cut into 1″ × 1″ and the titanium surface is adhered to the bottom of a 35 mm petri dish with a 13/16″ hole, using biologically inert silicone [9, 15]. In order to obtain powerful enough stress waves to initiate delamination, a waterglass layer was added to the back side of the samples, over the aluminum layer. Samples are then adhered to the bottom of a 35 mm diameter petri dish using a biologically inert silicone. Human blood plasma is then diluted down to 55% by volume in phosphate-buffered saline, to achieve the concentration found within the human body, and 1 ml is added on top of the titanium surface for 1 h. The solution of plasma is aspirated out of the dish leaving only the physiosorbed proteins on the titanium surface, Fig. 12.1a. Bacteria or cells are then inoculated inside of the substrate assembly until confluent films are formed.

Multiple substrate assemblies of biofilms constructed and tested, each film consisting of multiple loading locations. Onset of failure of the bacterial biomaterial-titanium interface is marked by spallation of the biofilm from the surface. For example, Fig. 12.1c depicts the adhesion failure progression for *S. mutans* on plasma coated roughened titanium substrates. As fluence, energy per area, values increase the spallation increases for the films tested. During testing the failure statistics for each film at each fluence are recorded in order to determine the minimal energy needed to initiate spallation. Uncultured smooth substrates are used as calibration specimen and are loaded with identical fluences used in the spallation study, and the free surface displacement is measured using a Michelson interferometer. The substrate stress profile can be measured using the free surface

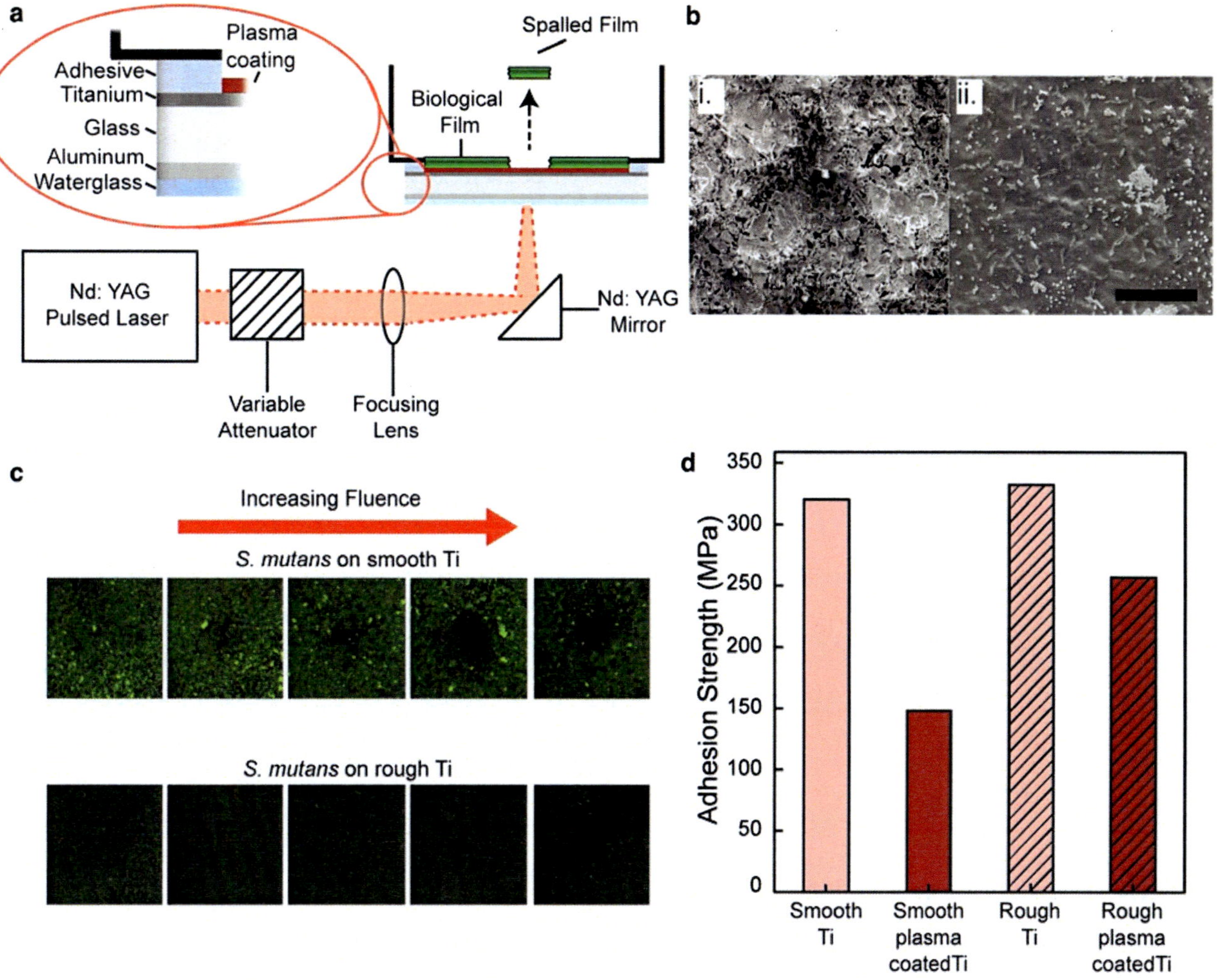

Fig. 12.1 (**a**) The laser spallation set up used during delamination experiments. The Nd: YAG laser used reflected 90° to allow for horizontal orientation of the substrate assembly used. (**b**) (i) Depicts an SEM image obtained of roughened titanium substrate used to mimic dental implant thread roughness, (ii) is an SEM image of the plasma protein coating on top of the titanium surface, scale bar is set to 100 μm. (**c**) Illustrate typical failure for both *S. mutans* plasma coated smooth and rough titanium surfaces due to increasing fluence values. (**d**) Graph of interface strength obtained after calibration experiments and Weibull analysis, for smooth and rough titanium and plasma coated smooth and rough titanium for *S. mutans*

velocity and converted into interface strength using previously accepted methods [19]. Smooth substrates of this type have previously been shown to be viable substitutes for the roughened sample as well due to the negligible effects of substrate stress propagation from the roughened surface [20]. Weibull analysis [21, 22], common in macroscopic adhesion analyses, calculates the half-life from a Weibull distribution, which is used as the adhesion strength. A decrease in adhesion strength is initially noted for the *S. mutans* films cultured on top of the plasma protein layer. Adhesion strength for *S. mutans* on smooth plasma coated titanium was measured as 148 MPa, and the adhesion strength measured on rough plasma coated titanium was 257 MPa. When we compare these values to previously uncoated smooth and rough titanium adhesion strength values, we can see that the plasma coating resulted in a much less strongly adherent biological film. Previous values for *S. mutans* on smooth and rough titanium are 320 MPa, with a 95% C.I. (304, 333), and 332 MPa, with a 95% C.I. (324, 343) respectively. There is a two-fold difference in adhesion strength between MG 63 monolayers and *S. mutans* biofilms on smooth titanium. Adhesion strengths values are shown in Fig. 12.1d. The adhesion strength for bacterial biofilms is greater than that of the cell monolayers. Future studies aim to examine the effect of plasma coated surfaces on cellular adhesion. Thus, resulting in a modified Adhesion Index. If cell adhesion remains constant or increases, then the resulting Adhesion Index will also increase. Meaning that in a more accurate in vivo model that surface roughness has a much more positive impact on adhesion strength for host cells than the bacterial adhesion.

This study directly compares the adhesion for *S. mutans* biofilms onto both smooth and roughened titanium dental implant mimicking surfaces, with and without a plasma pretreatment. Laser spallation was used in order to quantify the adhesion strength for the biological films. The adhesion values were directly compared to determine the effect of surface modifications and plasma protein coating on their respective adhesion. The inclusion of the plasma protein physiosorbed layer resulted in a decrease in adhesion for *S. mutans* on both the smooth and roughened titanium surfaces. Future work intends to further model in vivo conditions by understanding the effect of the plasma coated surface on MG 63 adhesion. Additionally, understanding the specific plasma proteins and their impact on adhesion is of interest. Fibronectin shall be applied to the titanium surfaces and more laser spallation experiments will be performed to understand its effect on adhesion. Furthermore, studies involving other implantable devices, like orthopedic implants, will be examined by varying the bacterial model to include invasive *Staphylococcus aureus*. The goal of this research is to expand upon an adhesion testing metric, the Adhesion Index. This Index will be used to aid in the design of medical devices, and ultimately be used to reduce biofilm-related infections while promoting the successful integration of these medical devices.

Acknowledgments We thank the Center for Pharmaceutical Research and Innovation (CPRI) for use of bacterial culture equipment. CPRI is supported, in part, by the University of Kentucky College of Pharmacy and Center for Clinical and Translational Science (UL1TR001998). We thank Drs. Larissa Ponomareva and Natalia Korotkova for sharing their bacterial culture expertise.

References

1. Velnar, T., Bailey, T., Smrkolj, V.: The wound healing process: an overview of the cellular and molecular mechanisms. J. Int. Med. Res. **37**(5), 1528–1542 (2009)
2. Park, J.B.: Biomaterials Science and Engineering. Springer (2012)
3. Barbosa, J.O., et al.: Streptococcus mutans can modulate biofilm formation and attenuate the virulence of Candida albicans. PLoS One. **11**(3), e0150457 (2016)
4. Kreth, J., Merritt, J., Qi, F.: Bacterial and host interactions of oral streptococci. DNA Cell Biol. **28**(8), 397–403 (2009)
5. Chang, Y.-Y., et al.: Biological characteristics of the MG-63 human osteosarcoma cells on composite tantalum carbide/amorphous carbon films. PLoS One. **9**(4), e95590–e95590 (2014)
6. Shapira, L., Halabi, A.: Behavior of two osteoblast-like cell lines cultured on machined or rough titanium surfaces. Clin. Oral Implants Res. **20** (1), 50–55 (2009)
7. Oshida, Y.: Bioscience and Bioengineering of Titanium Material, 2nd edn. Elsevier (2013)
8. Gaviria, L., et al.: Current trends in dental implants. J. Korean Assoc. Oral Maxillofac. Surg. **40**(2), 50–60 (2014)
9. Boyd, J.D., et al.: Biofilm and cell adhesion strength on dental implant surfaces via the laser spallation technique. Dent. Mater. **37**(1), 48–59 (2021)
10. Duarte, P.M., et al.: Bacterial adhesion on smooth and rough titanium surfaces after treatment with different instruments. J. Periodontol. **80**(11), 1824–1832 (2009)
11. Bayoudh, S., et al.: Assessing bacterial adhesion using DLVO and XDLVO theories and the jet impingement technique. Colloids Surf. B: Biointerfaces. **73**(1), 1–9 (2009)
12. Stoodley, P., et al.: Biofilm material properties as related to shear-induced deformation and detachment phenomena. J. Ind. Microbiol. Biotechnol. **29**(6), 361–367 (2002)
13. Grady, M.E., et al.: Molecular tailoring of interfacial failure. Langmuir. **30**(37), 11096–11102 (2014)
14. Wang, J.L., Weaver, R.L., Sottos, N.R.: A parametric study of laser induced thin film spallation. Exp. Mech. **42**(1), 74–83 (2002)
15. Boyd, J.D., Korotkova, N., Grady, M.E.: Adhesion of biofilms on titanium measured by laser-induced spallation. Exp. Mech. **59**(9), 1275–1284 (2019)
16. Hu, L., et al.: Cell adhesion measurement by laser-induced stress waves. J. Appl. Phys. **100**(8), 084701 (2006)
17. Hagerman, E., et al.: Evaluation of laser spallation as a technique for measurement of cell adhesion strength. J. Biomed. Mater. Res. A. **82**(4), 852–860 (2007)
18. Kearns, K.L., Boyd, J.D., Grady, M.E.: Biofilm rupture by laser-induced stress waves increases with loading amplitude, independent of location. ACS Appl. Biomater. (2019)
19. Kandula, S.S.V., et al.: Adhesion strength measurement of polymer dielectric interfaces using laser spallation technique. Thin Solid Films. **516** (21), 7627–7635 (2008)
20. Boyd, J.D., Grady, M.E.: The effect of surface roughness on laser-induced stress wave propagation. Appl. Phys. Lett. **117**(12), 121601 (2020)
21. Mei, L., et al.: Influence of surface roughness on streptococcal adhesion forces to composite resins. Dent. Mater. **27**(8), 770–778 (2011)
22. van der Mei, H.C., de Vries, J., Busscher, H.J.: Weibull analyses of bacterial interaction forces measured using AFM. Colloids Surf. B: Biointerfaces. **78**(2), 372–375 (2010)

Chapter 13
Mass Mitigation in Structural Designs via Dynamic Properties

J. Brent Knight

Abstract The efforts reported here are focused on mitigating unnecessary mass in aerospace hardware. The approach to remove this undesired mass from the design is to leverage both the dynamic strength of materials and the frequency dependency of strain. Analytically predicted dynamic responses of structures are often applied as static loads in stress analyses that ultimately dictate the weight of a structural design. Assuming a dynamic response is a static load and then comparing resulting stress predictions to a static strength property is a long-standing engineering practice. Doing so is known to be, or is assumed to be, conservative. However, little indication of the order of magnitude of embedded conservatism has been identified. NASA/MSFC efforts in 2011, 2019, 2020, and now in 2021 have begun to qualitatively show the order of magnitude of that conservatism. A quick turnaround engineering method is pursued to leverage the subject facets of physics for the purpose of decreasing the weight of flight hardware. Tests performed using simple beams and significant observations are described.

Key words Conservative design loads · Dynamic properties · Undue conservatism · Mass mitigation · Optimized structural design

13.1 Introduction

It is common, possibly almost exclusive, to use dynamic loads and environments to develop a structural design load that is applied as if it were static within the design/analysis process. Materials respond differently when subjected to dynamic loads as compared to static loads so this paradigm, assuming a dynamic load is static, has no physics basis. However, With Respect To (WRT) common ductile materials used in hardware designed for space applications, doing so is conservative. No work to quantify or qualify the amount of conservatism this introduces into the design has been identified. Therefore, one cannot determine how much this assumption costs a project or how much it limits Launch Vehicle (LV) or payload performance.

NASA MSFC has exerted efforts under its internal Technical Excellence (TE) program in 2011, 2019, and 2020 to quantify or qualitatively characterize the order of magnitude of conservatism imbedded in this engineering paradigm. In other words, how much unnecessary mass is in the design due to the subject assumption? If significant, perhaps it is worth investing resources to evolve a new engineering paradigm that leverages dynamic strength of materials and/or the frequency dependency of strain to facilitate lighter space flight hardware.

Highlights from 2011 efforts were presented at the Spacecraft and Launch Vehicle Dynamics working Group (SCLV) in 2012 [1]. The objective of this paper is to publish significant observations from all efforts to date.

J. B. Knight (✉)
NASA/MSFC, Huntsville, AL, USA
e-mail: brent.knight@nasa.gov

A. Amirkhizi et al. (eds.), *Challenges in Mechanics of Time Dependent Materials, Mechanics of Biological Systems and Materials & Micro-and Nanomechanics, Volume 2*, Conference Proceedings of the Society for Experimental Mechanics Series,
https://doi.org/10.1007/978-3-030-86737-9_13

13.2 Motivation

Ultimately, the motivation for this study is to develop a business case for funding to develop and evolve a new quick turnaround method to design hardware for dynamic environments that results in lighter hardware without necessitating development of new expensive "high performance" materials. Impacts or effects associated with assuming dynamic loads are static include:

1. Prevents optimizing structural designs WRT mass.
   (a) If there is significant undue conservatism in assuming a dynamic load is static then even with successful use of the state of the art optimization algorithms, hardware can still be made notably lighter.
2. Components and Secondary Structures (CSS), small items that attach to a LV, are sometimes notably heavier than need be due to design loads based on higher frequency accelerations.
3. Independent of the CSS mass, if the methodology used to assess the Primary Structure to CSS interface assumes dynamic loads are static and that includes unnecessary conservatism then that interface is more robust than need be. This can become an expensive engineering activity if the interface in question is a PS orthogrid or isogrid node. It should be extremely rare, if ever, that PS is sized to accommodate CSS.
4. PS itself is designed, in part, to dynamic loads. Therefore, there is undue conservatism/mass in PS designs that could be removed.
5. A given LV has limited uplift capability. If a payload is designed per the current method, that may limit the LVs that could transport that payload to space, since the payload would be heavier than need be.

Being able to leverage the subject facets of physics will enable a LV including all its subsystems being lighter which will allow the LV to lift heavier payloads. The same to be determined method will facilitate the payloads themselves being lighter. Both effects add, resulting in a much more efficient industry.

The weight of hardware is dependent on the load that it was designed to sustain. Therefore, it is important that no unnecessary component of load be imbedded in the resultant design load. Focus on perceived overly conservative design loads and on developing methods to decrease high loads is longstanding. However, the manner in which loads are applied in the stress analysis process is equally important WRT the final design weight. Little, if any, work has been done to show how much the subject assumption is costing a project has been done. *With it demonstrated that there is significant gain to be realized, funding to develop a new or modified engineering method will be pursued. That method will include measurement/development of frequency dependent material strength properties and a to be determined criteria that will facilitate defining a frequency threshold, for a given hardware item and specified environment, above which oscillating loads can be omitted in the structural design load development process.*

13.3 Background

Within the design phase of a new Launch Vehicle such as NASA's Space Launch system (SLS), efforts are exerted to design hardware within defined structural mass specifications. However, it is common for the mass of a given piece of hardware to be greater than need be due to schedule. If an item is designed and it has positive margins of safety it is likely more often than not signed off and engineering moves on to the next task. In many cases, mass could have been whittled away from that design but in that moment, meeting schedules is more important.

Methods exist to design for dynamic environments that are closer to physics based as compared to applying a derived net Center of Gravity (CG) acceleration as a static inertial load in a stress analysis, and they may provide mass relief. Those methods are likely more laborious/time consuming than the current approach. They also, in cases, necessitate additional testing which takes time.

Anecdotally speaking, minimal mass is very important in space flight hardware but in a fast and furious design project, "*schedule is king!*" For this reason, proposed new engineering methods have to be quick turnaround methods comparable, WRT required labor, to the current.

The current vision for a first cut modified method is to develop frequency dependent strength properties for materials commonly used in space hardware. That property would be used to assess predicted dynamic stresses. Also, criteria to be used to deem accelerations above a determined frequency threshold for a given hardware item negligible are to be developed. These envisioned methods are to directly plug and play into the current engineering flow so little to no schedule impact is perceived.

Fig. 13.1 Free vibration test articles and setup

Fig. 13.2 Forced vibration test article and setup

13.4 Tests

Two types of tests have been performed to qualitatively assess the order of magnitude of conservatism in a structural design that was based on applying a dynamic load statically. In both cases, simple aluminum beams were used as the Test Article (TA). In the first test type, "Free Vibration Test" (FrVT), the TA is a cantilever beam attached to a shaker table and in the second, "Forced Vibration Test" (FoVT), one end of the beam is fixed and a shaker is used to apply a specified sinusoidal force to the other end. Two FrVT graphics are shown in Fig. 13.1 and one FoVT graphic is shown in Fig. 13.2. There are two configurations associated with the 2011 activity. One is the simple aluminum beam alone and the other includes a large mass on the end of that beam to shift the first mode to a much lower frequency.

The FrVT configuration simulate a component mounted to (cantilevered) a LV wall. The FoVT configuration provides measured input forces to the TA.

13.4.1 Significant Observations: 2011 TE Efforts

Figure 13.1 presents the TA and the test set setup's two FrVT configurations. The intent of these tests was to measure strain from two similar tests. One with a "low" frequency fundamental mode and the other with a "high" frequency fundamental mode and with close to an equivalent loading condition. The metric of interest was how much strain would vary due to the frequency of oscillation if all other pertinent parameters were held equal to the degree possible. In both cases, the beam/cross section reacting the load was identical. The acceleration input was determined by specifying that the effective response

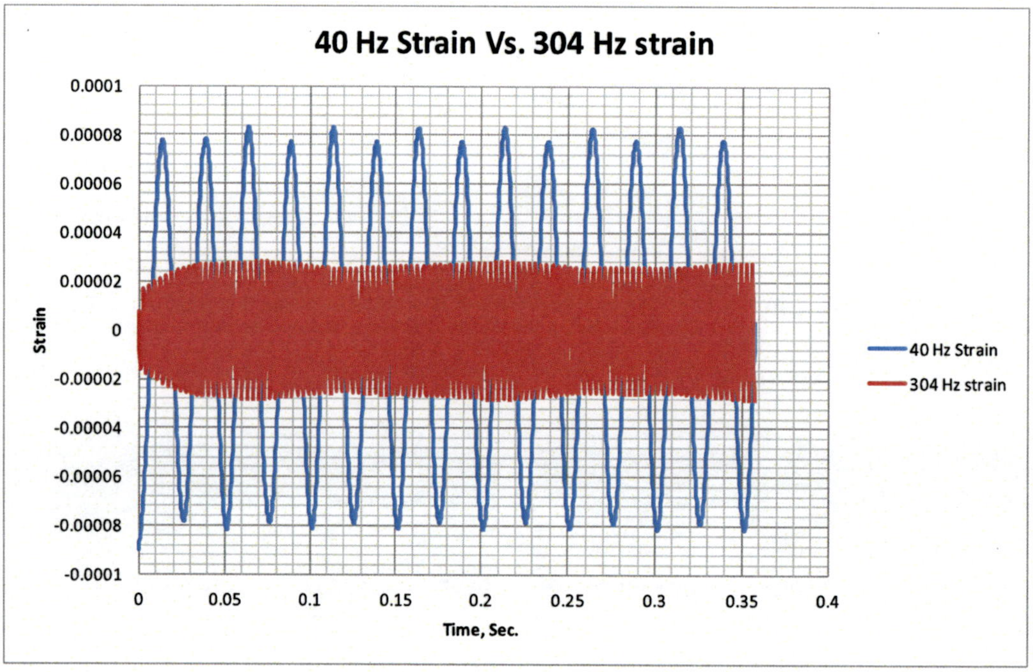

Fig. 13.3 2011 TE results

Table 13.1 Summary of 2011TE results

Configuration	Estimated force, Lb.	Acceleration, g	TA weight, Lb.	Peak strain	Fundamental mode, Hz[a]
1	98	24.5	4	0.00003	300
2	92	2	46	0.000084	40

[a](1) Approximate fundamental frequencies. (2) Input frequency

oscillating force for both cases be approximately equal and dividing by the mass in each case. The force was about 95 Lb. The tip response for configuration one was ≈24.5 g and that of configuration two was ≈2 g. Figure 13.3 shows the 40 and 304 Hz measured strain due to approximately the same load. *The low frequency strain was approximately a factor of three greater than the high frequency strain.*

This exhibits the "frequency dependency of strain" and that higher frequency motion, relatively speaking, can be expected to result in low strain. That being the case, strain associated with zero Hz is expected to be significantly higher than that associated with oscillating loads. Along these same lines, "low" frequency loads are more detrimental than "high frequency loads of the same amplitude. *All g's are not created equal!*

All of this equates to applying dynamic loads statically in stress analyses is conservative and as the frequency of the effective oscillating loads goes up the conservatism goes up as well. Configuration details and results are presented in Table 13.1.

13.4.2 Significant Observations: 2019 TE Efforts

Prior to dynamic testing, a static test was performed. Weights were suspended from the TA as shown in Fig. 13.4. The load necessary to reach the alloy's ultimate strength, about 80 Lb., was applied and a slight plastic deformation was observed. Figure 13.5 presents the post static test TA. The displacement of the TA tip while loaded was approximately 1″ which is consistent with predictions.

Like the 2011 efforts, 2019 efforts utilized the FrVT configuration but the objective was different. The TA and test configuration are shown in Fig. 13.1. Both sine dwell and random vibration tests were performed. The objective was to excite the first few modes of the TA and attempt to identify trends in measured strain WRT frequency of oscillation. The bulk of test results have not been post processed and summarized but one significant observation will be described.

Fig. 13.4 Static test

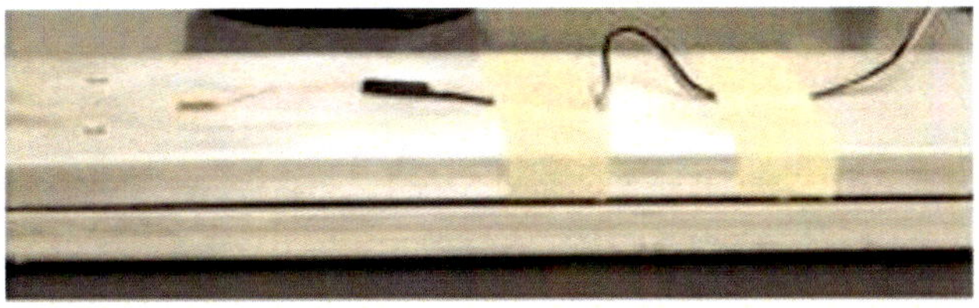

Fig. 13.5 Plastic deformation

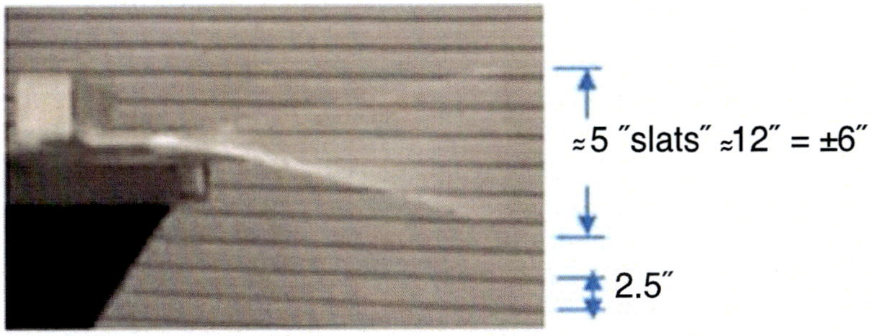

Fig. 13.6 Tip displacement

At the conclusion of the planned tests, the test engineer was asked to perform a final run and input the maximum amplitude acceleration that he was comfortable with and to do so at the fundamental mode of the TA. The response accelerometer was removed for its protection and 15 g was input at 43 Hz, the TA's first mode. Strain was measured in the test but about half way thru the test the strain gage failed, presumably due to the very large displacements of the TA. Figure 13.6 illustrates the magnitude of displacement that the TA experienced. An estimated ±6″ displacement was observed.

Comparing to the static test, approximately 1″ of flexure due to a static load resulted plastic deformation but ± 6″ of flexure dynamically did not. This alone illuminates the order of magnitude of conservatism embedded in assuming a dynamic load is static. Adding to that, the strain measured prior to strain gage failure was greater than twice the ultimate.

Fig. 13.7 TSS

Fig. 13.8 FoVT TA Constraints

13.4.3 Significant Observations: 2020 TE Efforts

Unlike 2011 and 2019 efforts, the 2020 efforts utilized the FoVT configuration shown in Fig. 13.2. These tests were performed to enable measurement of applied dynamic force directly. The primary target product from these tests was data that will support saying X pounds were applied dynamically to the TA without damage. Again, this rolls into qualitatively capturing the order of magnitude of conservatism in assuming a load is static in stress analyses.

The complete set of data from these tests has not been processed and summarized. The single most significant observation is described here. With one end of the TA fixed and the other engaged with the Test Support Structure (TSS) in a kinematic manner (situated between bearings in the TSS), the shaker table applied the dynamics load. The TSS included three load cells that separated the two steel plates and the test was controlled per the average of those measured forces. Figure 13.7 shows the TSS. Figure 13.8 shows the TA's constraints.

An applied force at specified amplitudes and frequencies was applied to the tip (approximately 0.5″ from the end) of the beam. *The most obvious significant finding was that at 40 Hz, categorically low frequency, the TA sustained 320 Lb. which is 4 times the static load that resulted in plastic deformation.*

Once again, findings make clear that there is significant gain to be realized via evolving a quick turn around stress analysis method that leverage the subject facets of physics.

13.5 Future Work

The work presented in this paper was relative to aluminum beams. Future efforts are planned to investigate different alloys and perhaps various composites. Additionally, efforts to assess joints are planned as well. Threaded fasteners, inserts, and welds are to be tested.

13.6 Conclusion

It has been concluded that there is significant undue conservatism in assuming dynamic loads are static in stress analyses that dictate the weight of hardware designed for space applications. When the envisioned dynamic strength properties and criteria are in hand, notable mass savings will be realized. Intuition suggest that similar results and conclusions will be arrived at for other alloys and possibly some composites.

Prior to implementation, similar tests are to be performed to assess bolted and welded joints.

Acknowledgments Efforts associated with the content of this paper were exerted as time permitted in between required project support. Over the course of years, numerous people and organizations have contributed to this effort. NASA/MSFC has provided resources on three occasions. NASA/MSFC/ET40 (Structural Dynamic Test Lab) and specifically Mr. Steve Rodgers, has provided support as time was available. MSFC management has proactively assisted in efforts to get support for this activity. Numerous interns have contributed much. The interns that supported are Rachel Pilgrim (spring 2019), Ashlee Bracewell (summer 2019), Miguel Alizo (Fall 2019), Marcio Simao (spring and summer 2020), and Ariel Ferrera (summer 2020).

Reference

1. Knight, J.B.: Undue Conservatism in Random Vibration Design Loads. SCLV, Aerospace Corporation (2012)

Chapter 14
High-Temperature Burst Creep Properties of Nuclear-Grade FeCrAl Fuel Cladding

Abdullah S. Alomari, P. Joshi, and K. L. Murty

Abstract The cladding acts as the primary protection barrier for nuclear fuel in nuclear reactors. Following the aftermath of loss of coolant accident at Fukushima Daiichi plant in 2011, considerable international efforts have been directed toward developing newly Accident-Tolerant Fuel and cladding (ATF) materials for light water reactors (LWRs) as alternatives to zirconium-based alloys. The main purpose of ATF materials is to maintain the integrity of the core components under normal operation and to slow core degradation at a high-temperature steam environment under severe accident scenarios such as loss of coolant. Preliminary investigations suggest that Iron-Chromium-Aluminum (FeCrAl) based alloy system can be utilized for cladding applications in LWRs due to their excellent high-temperature steam oxidation, corrosion performance and high-temperature strength. To promote a better understanding about failure behavior, stress rupture tests for thin-walled tubing FeCrAl C26M2 grade have been conducted at temperatures ranging from 480 to 650 °C. The tubes were pressurized by inert Argon gas in the pressure range of 500–6200 psi (corresponding to hoop stresses of 170–376 MPa using the thin-tube approximation). Time to rupture, uniform strain, steady-state creep rate were determined and analyzed as functions of the temperatures and applied pressures. An apparent decrease of burst time was observed with increasing pressure/temperature. Using burst data of FeCrAl alloy obtained in this study, Larson–Miller Parameter and Monkman–Grant relationship were developed and compared with other ferritic alloys. Further, high temperature deformation behavior was found to obey the power-law creep with stress exponent of 5.2 ± 0.9 and an activation energy of 290 ± 32 kJ/mol. In addition, the FeCrAl tubing was found to fail by two different fracture modes: either by direct open-up or small crack and pinhole formation depending on applied temperature and pressure. The results are discussed and compared with commonly used cladding materials.

Key words Accident-tolerant fuel and cladding (ATF) · Light water reactors (LWRs) · Cladding burst · Monkman–Grant relationship · Power-law creep

14.1 Introduction

For several decades, Light Water Reactors (LWRs) have been extensively utilized to feed the electrical grid in more than 20 countries around the world [1]. The heat produced from the nuclear fuel in LWRs is transferred to the water surrounding the fuel cladding tubes to provide steam which is then employed to spin the turbines for electricity production. Thus, the fuel cladding isolates the radioactive materials generated within nuclear fuel from the water coolants and acts as the primary physical protection barrier. The selection of fuel cladding materials must satisfy various design constraints including low neutron absorption cross section, good mechanical properties under elevated temperatures and high radiation fields, adequate thermal conductivity and high corrosion resistivity to water [2]. Of course, there is a trade-off between multiples engineering requirements and the choice between different materials depends on attaining maximum desired properties. Zirconium-based alloys of various types have been historically selected as cladding materials in LWRs due to multiple favorable properties such as low cross-section for thermal neutron absorption, adequate strength and ductility, corrosion resistance to water and

A. S. Alomari (✉)
Nuclear Science Research Institute, King Abdulaziz City for Science and Technology, Riyadh, Saudi Arabia
e-mail: asalomari@kacst.edu.sa; asalomar@ncsu.edu

P. Joshi · K. L. Murty
Department of Nuclear Engineering, North Carolina State University, Raleigh, NC, USA
e-mail: pjoshi5@ncsu.edu; murty@ncsu.edu

A. Amirkhizi et al. (eds.), *Challenges in Mechanics of Time Dependent Materials, Mechanics of Biological Systems and Materials & Micro-and Nanomechanics, Volume 2*, Conference Proceedings of the Society for Experimental Mechanics Series,
https://doi.org/10.1007/978-3-030-86737-9_14

compatibility with fuel [2]. While Zircaloy-2 and Zircaloy-4 alloys have continued to be the backbone fuel-cladding materials of LWRs for several decades, further development in mechanical and corrosion properties of zirconium alloys has also been attained such as Nb-containing Zirlo and M5 alloys [3].

However, a main disadvantage of zirconium alloys is the severe degradation of cladding properties at very high temperatures beyond design basis accident. In case of loss of coolant accident (LOCA), the water coolant is no longer in liquid condition and cladding is surrounded by high temperature steam which initiates an exothermic oxidation of zircaloy fuel cladding, resulting in substantial production of hydrogen which may lead to explosion and a direct threat to containment integrity. The aftermath of loss of coolant accident (LOCA) at Fukushima Daiichi nuclear power plant in 2011, caused by the most powerful earthquake ever recorded in Japan followed by strong tsunami, have triggered international efforts toward developing newly Accident-Tolerant Fuel and cladding (ATF) materials for light water reactors (LWRs) as alternatives to zirconium-based alloys. The main objective of ATF materials is to maintain the integrity of the core components under normal operation and to slow core degradation at a high-temperature steam environment under severe accident scenarios such as LOCA. Various promised ATF cladding materials and technologies have been nominated such as coated Zr-based cladding, Iron-Chromium-Aluminum (FeCrAl) based, and SiC/SiC cladding. Development status, potentials, challenges, and research needs for those candidate ATF materials have been recently explored [4, 5]. It has been shown that FeCrAl based alloy system can be utilized for cladding applications in LWRs due to their excellent high-temperature steam oxidation, corrosion performance and high-temperature strength. Multiple minor alloying additions of FeCrAl alloys with various heat treatments have been examined to arrive a better combination of thinner tube without losing mechanical and thermal properties. Among those, the wrought FeCrAl alloy C26M was down selected for the GenII LWRs [6]. Since the cladding tubes are subjected to hoop loading conditions due to internal pressure of the gas filled between the fuel and the cladding tube, it is of importance to examine the behavior of burst and thermal creep properties in order to predict the mechanical performance of the cladding during the lifetime of a reactor. In this work, burst rupture tests for thin-walled tubing of the nuclear-grade FeCrAl alloy (FeCrAl C26M2 grade), have been carried out at various temperatures and pressure levels to understand the burst behavior. The results are discussed and compared with other relevant cladding materials.

14.2 Experimental Methods

Chemical compositions of the nuclear-grade FeCrAl C26M2 tube investigated in this study is shown in Table 14.1. The length, inner diameter, and outer diameter of the tube samples were 12.7 cm, 9 mm, and 9.4 mm, respectively. A photograph of a typical burst specimen is shown in Fig. 14.1c. High temperature burst tests were conducted by pressurizing the tube with inert argon gas from a highly modified ATS® Series 1815 with sustained gas pressurization system. The gas pressurization unit and the burst tester chamber are shown in Fig. 14.1. The tubes were pressurized at constant pressure levels ranging of

Table 14.1 Chemical composition (wt. %) of the FeCrAl C26M employed in this study

Element	Cr	Al	Mo	Si	Y	C	S	Fe
(Wt. %)	11.87	6.22	1.98	0.2	0.03	<0.01	<0.005%	Bal.

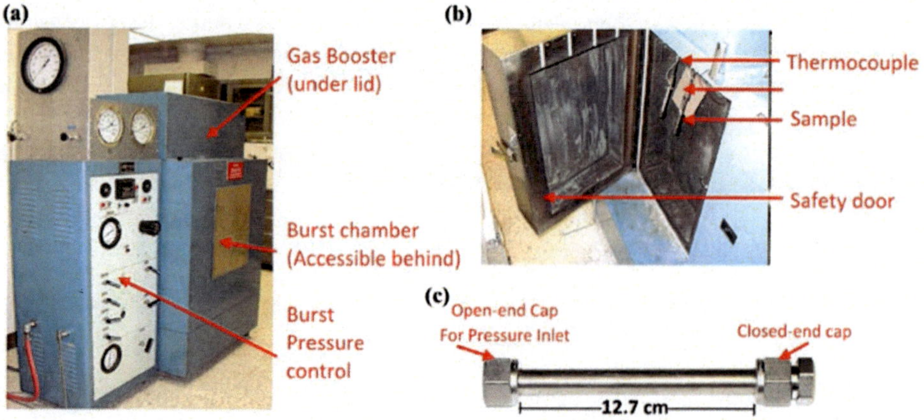

Fig. 14.1 Experimental setup: (**a**) Burst Tester showing the gas pressurization unit, (**b**) Burst Tester Chamber (**c**) A typical burst specimen

500–6200 psi at constant temperatures ranging from 480 to 650 °C. The temperatures were monitored using two K-type thermocouples attached to the specimen with an accuracy of ±5 °C. To minimize the gas volume during burst tests, a brass mandrel was inserted into the specimens before attaching closed-end cap so that damage to the experimental setup could be reduced. The time to rupture (burst) was then recorded using a high-speed camera attached to the pressure gauge. At a given internal gas pressure level (P), the hoop stress (σ_θ) of the FeCrAl tube was determined using the thin-tube approximation as follows

$$\sigma_\theta = \frac{PD_i}{2t} \tag{14.1}$$

where D_i is the inner diameter and t the tube thickness.

14.3 Results and Discussions

14.3.1 Burst Behavior

The dependence of burst rupture time, t_r on the applied hoop stress, σ_θ for FeCrAl C26M2 alloy at various temperatures is shown in Fig. 14.2 depicting linear relationship in a log–log scale. At all examined temperatures, an apparent decrease of burst rupture time was observed as hoop stress (pressure) or temperature increasing. It is interesting to compare the data obtained here in this study under constant hoop stresses and constant temperatures to those under LOCA accident experiment [7]. Figure 14.3 shows the burst temperature as a function of hoop stress for the FeCrAl C26M employed in this study in comparison with various cladding materials obtained from Pint et al. [8]. Given the same hoop stress and relatively similar alloy (the FeCrAl C26M2), the onset of burst for alloy here is at lower temperature than that obtained under LOCA accident experiment.This observation can be rationalized based on the effect of creep deformation during burst tests. Further details about the creep behavior of the FeCrAl C26M2 are discussed in the next section.

To characterize the interrelation between of the burst time on the applied hoop stress, the Larson–Miller parameter (LMP) was employed for the FeCrAl C26M2 alloy using the following expression,

$$\text{LPM} = T \cdot [C + \log(t_r)] \tag{14.2}$$

where T is in K, t_r in h and C is a constant equal to 20. Figure 14.4a shows the applied stress vs LMP at different temperatures. It is clearly seen that values of LMP are observed to be independent of the temperature at a given hoop stress where all data coalesce into a single line. Furthermore, the steady state creep rate, $\dot{\varepsilon}_{ss}$ was estimated from the variation in the tube outer diameter as [9],

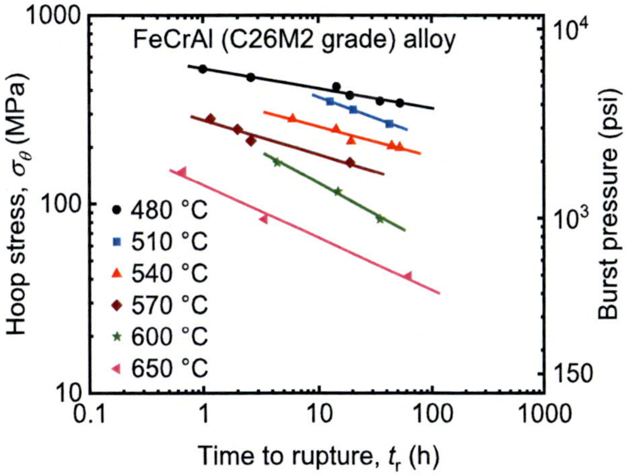

Fig. 14.2 Log–log plot of the hoop stress versus time to rupture (burst) at various temperatures of the nuclear-grade FeCrAl alloy

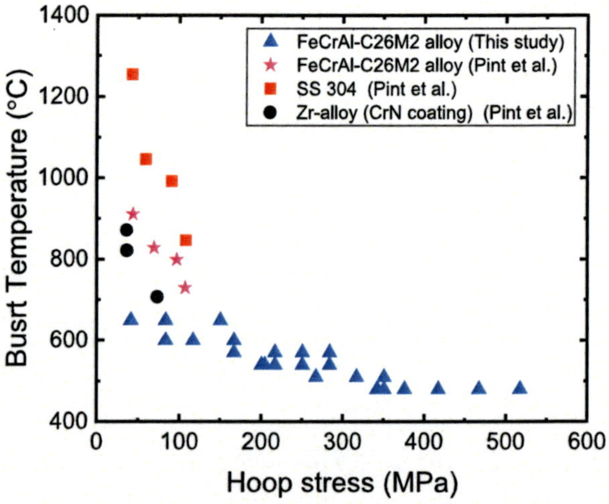

Fig. 14.3 Burst temperature as a function of hoop stress for various cladding materials for the FeCrAl C26M employed in this study in comparison with various cladding materials [8]

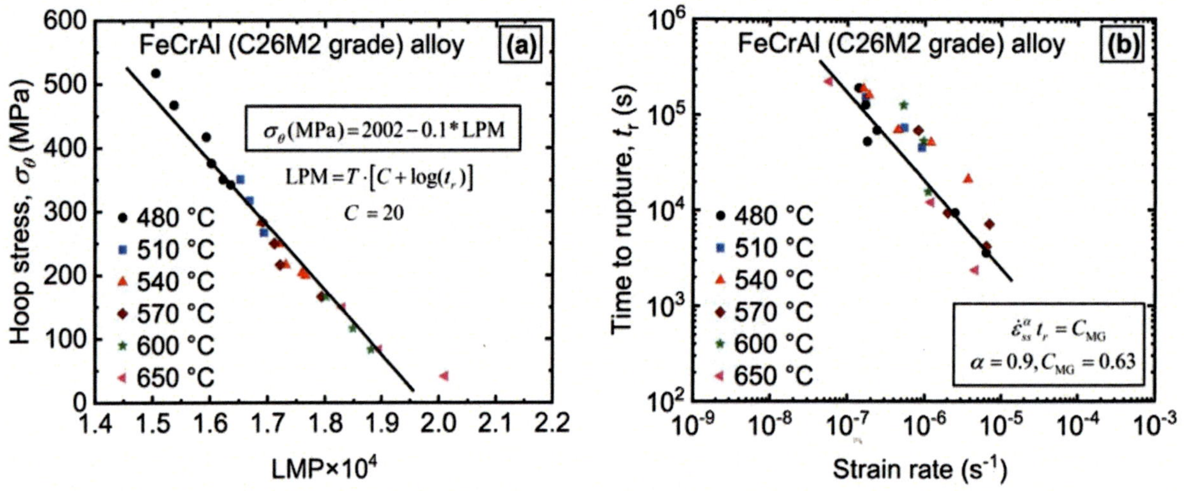

Fig.14.4 (**a**) Larson–Miller parameter (LMP) plot at different temperatures for the nuclear-grade FeCrAl alloy. (**b**) Log–log plot of the rupture life vs steady-state creep rate exhibiting Monkman–Grant relation

$$\dot{\varepsilon}_{ss} = \frac{\ln\left(\frac{D}{D_o}\right)}{t_r} \tag{14.3}$$

where D is the average outer diameter after burst tests measured away from the burst region and D_o is the original diameter of the specimen before test.

The variation of time to burst, t_r as a function of the steady state creep rate at various temperatures for the FeCrAl C26M2 alloy is shown Fig. 14.4b. The alloy is found to follow the Monkman–Grant relation expressed as,

$$\dot{\varepsilon}_{ss}^{\alpha} \cdot t_f = C_{MG} \tag{14.4}$$

where α and C_{MG} are constants 0.9 and 0.63, respectively.

The fractured samples after bursting were generally found to follow two different fracture modes: either by direct open-up or small crack and pinhole formation depending on applied temperature and pressure. Typical photographs for the nuclear-grade FeCrAl alloy showing the two different modes of failure are shown in Fig. 14.5. In general, the larger the internal hoop stresses resulting from cladding overpressure, the most likely the sample fails by direct burst opening while burst by a pinhole formation in the ballooning region observed at lower stresses.

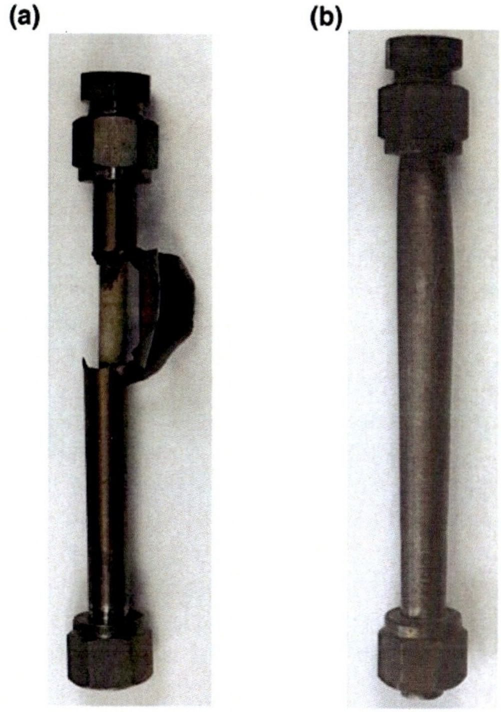

Fig. 14.5 Typical photographs for the nuclear-grade FeCrAl alloy after the burst test showing the two different fracture modes (**a**) direct open-up and (**b**) small crack and pinhole formation

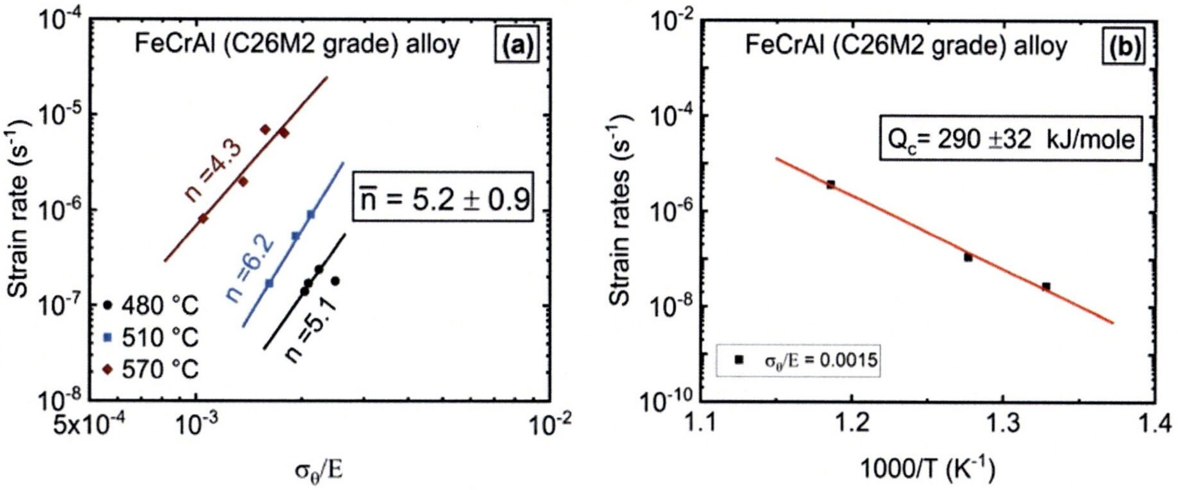

Fig. 14.6 (**a**) Log–log plot of the steady state creep rate versus normalized hoop stress at various temperatures for the nuclear-grade FeCrAl alloy. (**b**) Semilog plot of the steady state creep rate versus reciprocal temperature at a normalized hoop stress (σ/E) of 0.0015

14.3.2 High Temperature Creep Behavior

Figure 14.6a shows log–log plot of the steady state creep rate determined form Eq. 14.3 as function of the normalized hoop stress at various temperatures for the nuclear-grade FeCrAl (C26M2 grade) alloy. The elastic modulus of the FeCrAl alloy was estimated using the following empirical formula [10],

$$E = -5.46 \times 10^{-5}T^2 - 3.85 \times 10^{-2}T + 199 \tag{14.5}$$

where E is the elastic modulus in GPa and T is temperature in °C.

The steady-state creep rate is found to be related to the hoop stress through a power-law relation,

$$\dot{\varepsilon}_{ss} = A\left(\frac{\sigma_\theta}{E}\right)^n \exp\left(\frac{-Q}{RT}\right) \tag{14.6}$$

where A is a constant and n the stress exponent, Q the activation energy for creep, T temperature and R gas constant. The average stress exponent was found to be 5.2 ± 0.9 obtained from the Fig. 14.6a. To estimate the activation energy for creep, an Arrhenius plot of the $\dot{\varepsilon}_{ss}$ vs reciprocal of the absolute temperature at a constant normalized hoop stress is shown in Fig. 14.6b where Q for creep deformation was found to be 290 ± 32 kJ/mol.

The values for the stress exponent and the activation energy for creep were commonly employed as indicative parameters to infer the micro-mechanism(s) controlling creep deformation. For example, when the stress exponent yields a value of 3–3.5 and the activation energy being close to that of the solute diffusion, viscous glide process is controlling the creep deformation while thermally activated dislocation climb process becomes the rate-controlling mechanism for higher values of n ranging from 4.5 to 7 and activation energies close to that of the lattice self-diffusion. In this study, the obtained value of the activation energy for the FeCrAl (C26M2 grade) alloy (290 kJ/mol) was found to be comparable to of that of self-diffusion of Fe in alpha-structure (251 kJ/mol). Accompanied with the obtained value of the stress exponent ($n \sim 5$), dislocation climb is considered to be the rate-controlling creep mechanism in FeCrAl (C26M2 grade) alloy at examined temperatures 480–650 °C. This conclusion was further supported by the observation of well-defined sub-grains with their boundaries stacked with dislocations as typical reflections for the dislocation climb mechanism [11, 12].

14.4 Summary and Conclusion

To understand the burst behavior of the nuclear-grade FeCrAl alloy (FeCrAl C26M2 grade), burst rupture tests for thin-walled tubing have been carried out at temperatures ranging from 480 to 650 °C and at pressure levels (using inert Argon gas) ranging from 500 to 6200 psi (corresponding to hoop stresses of 170–376 MPa). The following conclusions have been made,

- Apparent decrease of burst time of the FeCrAl alloy was observed as pressure or temperature increasing.
- Larson–Miller Parameter and Monkman–Grant relationship were developed and compared with other ferritic alloys.
- High temperature creep deformation behavior was found to obey the power-law creep with a stress exponent of 5.2 ± 0.9 and an activation energy of 290 ± 32 kJ/mol.
- According to the obtained values of the stress exponent and the activation energy, dislocation climb is considered to be the rate-controlling creep mechanism during deformation.
- The FeCrAl tubing was found to fail by two different fracture modes: either by direct open-up or small crack and pinhole formation depending on applied temperature and pressure.

Acknowledgments Authors are thankful to Yukinoro Yamamoto from Oak Ridge National Laboratory for providing the specimens; Kurt Terrani and Andy Nelson for their support. This work was partially supported by the U.S. Department of Energy, Office of Nuclear Energy under DOE Idaho Operations Office Contract DE-AC07-051D14517. North Carolina State University personnel acknowledge the support of the National Science Foundation grant CMMI1727237.

References

1. Rebak, R.B.: Chapter 2—Current materials in light water reactors. Why do we need a materials renewal? In: Rebak, R.B. (ed.) Accident Tolerant Materials for Light Water Reactor Fuels, pp. 15–41. Elsevier (2020)
2. Murty, K.L., Charit, I.: An Introduction to Nuclear Materials: Fundamentals and Applications. Wiley-VCH (2013)
3. Charit, I.: Accident tolerant nuclear fuels and cladding materials. JOM. **70**(2), 173–175 (2018)
4. Chen, S.-L., He, X.-J., Yuan, C.-X.: Recent studies on potential accident-tolerant fuel-cladding systems in light water reactors. Nucl. Sci. Tech. **31**(3), 32 (2020)
5. Terrani, K.A.: Accident tolerant fuel cladding development: promise, status, and challenges. J. Nucl. Mater. **501**, 13–30 (2018)
6. Yamamoto, Y., Kane, K., Pint, B., Trofimov, A., Wang, H.: Report on Exploration of New FeCrAl Heat Variants with Improved Properties. Oak Ridge National Laboratory., M3FT-19OR020202053, ORNL/TM-2019/1290 (2019)
7. Massey, C.P., Terrani, K.A., Dryepondt, S.N., Pint, B.A.: Cladding burst behavior of Fe-based alloys under LOCA. J. Nucl. Mater. **470**, 128–138 (2016)
8. Pint, B.A., Baldesberger, L., Kane, K.: Steam Oxidation, Burst and Critical Heat Flux Testing of Commercial FeCrAl Cladding, United States, p. 2 (2019)

9. Srikant, G., Marple, B., Charit, I., Murty, K.L.: Characterization of stress rupture behavior of commercial-purity-Ti via burst testing. Mater. Sci. Eng. A. **463**(1), 203–207 (2007)
10. Thompson, Z.T., Terrani, K.A., Yamamoto, Y.: Elastic Modulus Measurement of ORNL ATF FeCrAl Alloys, United States (2015). https://doi.org/10.2172/1225433
11. Alomari, A.S., Kumar, N., Murty, K.L.: Creep behavior and microstructural evolution of a Fe-20Cr-25Ni (mass percent) austenitic stainless steel (alloy 709) at elevated temperatures. Metall. Mater. Trans. A. **50**(2), 641–654 (2019)
12. Joshi, P., Kombaiah, B., Cinbiz, M.N., Murty, K.L.: Characterization of stress-rupture behavior of nuclear-grade C26M2 FeCrAl alloy for accident-tolerant fuel cladding via burst testing. Mater. Sci. Eng. A. **791**, 139753 (2020)